QE
65
M3

Silurian and Lower Devonian
Basin and Basin-Slope Limestones
Copenhagen Canyon, Nevada

JONATHAN C. MATTI
MICHAEL A. MURPHY
STANLEY C. FINNEY

HILL
LIBRARY
MAY 2 8 1975
ST. PAUL

**THE
GEOLOGICAL SOCIETY
OF AMERICA**

SPECIAL PAPER 159

Silurian and Lower Devonian
Basin and Basin-Slope Limestones
Copenhagen Canyon, Nevada

JONATHAN C. MATTI
Department of Geology
Stanford University
Stanford, California 94305

MICHAEL A. MURPHY
Department of Earth Sciences
University of California, Riverside
Riverside, California 92502

STANLEY C. FINNEY
Department of Geology
Ohio State University
Columbus, Ohio 43210

THE
GEOLOGICAL SOCIETY
OF AMERICA

SPECIAL PAPER 159

WILLIAM MADISON RANDALL LIBRARY UNC AT WILMINGTON

Copyright 1975 by The Geological Society of America, Inc.
Library of Congress Catalog Card Number 74-19734
I.S.B.N. 0-8137-2159-8

Published by
THE GEOLOGICAL SOCIETY OF AMERICA, INC.
3300 Penrose Place
Boulder, Colorado 80301

Printed in the United States of America

QE665
.M43

The printing of this volume has been made possible through the bequest of Richard Alexander Fullerton Penrose, Jr., and the generous support of all contributors to the publication program.

245920   LIBRARY

Contents

Acknowledgments

M. O. Woodburne, K. M. Nichols, H. E. Cook, and F. G. Poole kindly read earlier drafts of this manuscript. J. G. Johnson read the manuscript and shared with us his ideas on Lower Devonian stratigraphy in central Nevada. The comments and criticisms of these geologists are appreciated. G. Klapper permitted us to use his detailed, unpublished conodont biostratigraphic data from the lower Kobeh Member of the McColley Canyon Formation at McColley Canyon.

The work was accomplished while we were affiliated with the Department of Earth Sciences, University of California, Riverside, and was supported by National Science Foundation Grants GA-1035 and GA-12736 to Murphy and the University of California, Riverside Intramural Research Fund.

Abstract

Silurian and Lower Devonian carbonate rocks in central Nevada are grouped into two distinct, time-equivalent assemblages: (1) an eastern dolomite suite, deposited on a broad, peritidal carbonate platform; and (2) a western limestone-clastic suite, deposited in subtidal, predominantly basinal environments oceanward from the margin of the dolomite suite platform.

Silurian through Lower Devonian (Llandoverian through Pragian) limestones belonging to the limestone-clastic suite occur at Copenhagen Canyon, Nevada. This sequence is conformable and is divided into three formations, which include, in ascending order, the Roberts Mountains Formation, the Windmill Limestone, and the Rabbit Hill Limestone.

The Silurian–Lower Devonian rocks are predominantly graptolitic, evenly laminated, argillaceous lime mudstone and wackestone interpreted as basin and basin-slope deposits. Graded and nongraded bioclastic limestone beds carrying fragmented shoal-water benthic faunas are interbedded with the mudstone-wackestone sequence. The bioclastic sediment is interpreted as allochthonous calcareous sand (allodapic packstone and grainstone) that was probably deposited by turbidity currents. Deposition by grain-flow and debris-flow mechanisms may also have occurred.

The Roberts Mountains Formation and the Windmill Limestone are interpreted as distal, level-bottom sequences. They are western equivalents of the Lone Mountain Dolomite, a pervasively dolomitized shoal-water carbonate complex that accumulated east of Copenhagen Canyon on the margin of the dolomite suite platform. Allodapic sand in the Roberts Mountains Formation and in the Windmill Limestone was derived from this eastern carbonate build-up.

The lower member of the Rabbit Hill Limestone is a distal, level-bottom sequence and is a western equivalent of the lower Kobeh Member of the McColley Canyon Formation. The upper member of the Rabbit Hill Limestone is a proximal sequence deposited on the basin slope, as indicated by thick allodapic sand and soft-sediment slump structures. Proximal basin-slope conditions in the upper member resulted from basinal progradation of the Kobeh Member carbonate-platform complex during deposition of the upper Rabbit Hill member. Thus, the upper Rabbit Hill member in the limestone-clastic suite is time-correlative with the lower Kobeh Member in the dolomite suite, but it was subsequently overlain by the prograding Kobeh Member.

Spinoplasia Zone brachiopods occur in the upper Rabbit Hill member. The *Spinoplasia* Zone and the *Trematospira* Subzone were previously thought to be entirely superpositional in relation. However, we believe the two brachiopod assemblages are facies-controlled, largely coeval benthic communities.

1

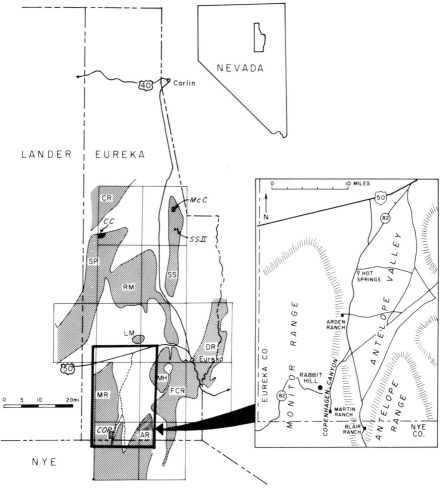

Figure 1.   Index map of localities referred to in this report. AR = Antelope Range; CC = Coal Canyon; COP = Copenhagen Canyon; CR = Cortez Range; DR = Diamond Range; FCR = Fish Creek Range; LM = Lone Mountain; McC = McColley Canyon; MH = Mahogany Hills; MR = Monitor Range; RM = Roberts Mountains; SP = Simpson Park Range; SSII = Measured section through upper Lone Mountain Dolomite and lower McColley Canyon Formation near Telegraph Canyon in the Sulphur Springs Range.

Introduction

This study synthesizes biostratigraphic and petrographic investigations of Silurian and Lower Devonian limestones exposed along the west flank of Copenhagen Canyon, southwestern Eureka County, Nevada (Fig. 1). The objective of this paper is threefold: (1) to describe the geology and stratigraphy of the Silurian–Lower Devonian carbonate strata, (2) to evaluate the depositional environments of these rock units, and (3) to propose time-stratigraphic correlations between various Lower Devonian rock units in central Nevada.

The well-exposed, fossiliferous, and depositionally continuous Silurian–Lower Devonian sequence at Copenhagen Canyon is important for two reasons. (1) Provincial shelly faunas and cosmopolitan graptolite and conodont faunas are interbedded in the sequence (Johnson, 1970, 1974; Finney, 1971; Matti, 1971). Copenhagen Canyon thus provides a valuable reference for correlation between different biofacies and between biogeographic provinces during Silurian and Early Devonian (Llandoverian through Pragian) time (J. C. Matti and M. A. Murphy, in prep.). (2) Copenhagen Canyon exposures of the Rabbit Hill Limestone clarify stratigraphic and depositional relations between the Rabbit Hill and other Lower Devonian rock units in central Nevada.

The Rabbit Hill Limestone has been an enigmatic unit in the Lower Devonian of central Nevada. Its exact stratigraphic position has been ambiguous because (1) the boundaries of the formation have never been adequately defined, (2) the lithology and biostratigraphy of the Rabbit Hill interval have never been described in detail, (3) a depositional model for the formation has never been proposed, (4) lithofacies and biofacies changes in central Nevada (discussed below) make Silurian and Lower Devonian lithologic and temporal correlation difficult, and (5) stratigraphic relations between middle Lower Devonian (Pragian) conodont, graptolite, and shelly faunas in central Nevada have not been worked out in detail. For these reasons, contrasting stratigraphic models have been proposed for the Rabbit Hill Limestone (Johnson, 1965, 1970; Clark and Ethington, 1966; Matti, 1971; Merriam, 1973).

In this paper we attempt to clarify stratigraphic relations of the Rabbit Hill Limestone by (1) describing the lithology and physical stratigraphy of Silurian–Lower Devonian rock units at Copenhagen Canyon, (2) defining a mappable lower contact for the Rabbit Hill Limestone in its type area, (3) proposing a depositional model for the Rabbit Hill, (4) outlining the biostratigraphy of the Rabbit Hill Limestone in its type area, and (5) proposing that the Rabbit Hill Limestone in the limestone-clastic suite is largely time-correlative with lower portions of the Kobeh Member of the McColley Canyon Formation in the dolomite suite. This correlation clarifies stratigraphic and depositional relations between the dolomite and limestone-clastic suites during Rabbit Hill (Pragian) time.

3

PREVIOUS INVESTIGATIONS

Figure 2 summarizes pertinent stratigraphic studies at Copenhagen Canyon and the evolution of Silurian-Devonian stratigraphic nomenclature in central Nevada.
In a reconnaissance survey of the Roberts Mountains region, Merriam and Anderson (1942, p. 1687) cited a Middle Silurian (Clinton) age for lower beds of the Roberts Mountains Formation at Copenhagen Canyon and first drew attention to "Silurian" brachiopod-bearing limestone strata, which Merriam (1954) later correlated with the Lower Devonian (Helderbergian) of the New York succession. Merriam (1963) subsequently presented a geological report on the Antelope Valley area in which he described the general geology of Copenhagen Canyon and introduced the name "Rabbit Hill Limestone" for beds bearing the Helderbergian brachiopods. Merriam (1973) later described the taxonomy of shelly fossils in the Rabbit Hill Limestone.
More recent investigations have focused on the rich Copenhagen Canyon Lower Devonian faunal succession. On the basis of reconnaissance collections, Clark and Ethington (1966) compared conodonts from the type section of the Rabbit Hill Limestone with "Siegenian?-Emsian" faunas from Frankenwald, Germany. Johnson (1965, 1970, 1974) studied brachiopod faunas from the Rabbit Hill Limestone at Copenhagen Canyon and at Coal Canyon in the Simpson Park Range. On the basis of these collections he (1965) defined the *Spinoplasia* Assemblage Zone. Berry and Murphy (1972) reported the occurrence of *Monograptus thomasi* and *M. yukonensis* from the type section of the Rabbit Hill Limestone. Finney (1971) and Matti (1971) conducted a detailed biostratigraphic study of Lower Devonian graptolite and conodont faunas at Copenhagen Canyon, and Matti and Murphy have studied the petrography of the Silurian–Lower Devonian rock units. These joint investigations are the basis of the work presented here.

TERMINOLOGY

Allochthonous. This term refers to material whose dominant constituents were not formed in place but were transported from their original habitat or depositional site into a significantly different depositional environment.
Allodapic. Meischner (1964, p. 173) introduced the term "allodapic" for allochthonous calcareous sand carrying shoal-water faunas that is interbedded with pelitic basinal sediment. At Copenhagen Canyon the dominant allodapic constituents are skeletal and nonskeletal material derived from contemporaneous shoal-water carbonate environments. Meischner proposed that allodapic limestone beds are deposited from turbidity currents. We believe that most allochthonous sands at Copenhagen Canyon were also deposited by turbidity currents. Hence, "allodapic" is mainly a genetic term, although we also use "allodapic" descriptively as a concise term for skeletal and nonskeletal packstone and grainstone that may be graded, laminated, cross-laminated, or massive, and that may fine upward into mudstone and wackestone host rocks.
Bank. "A deposit which is composed of essentially in-place skeletal and non-skeletal carbonate material which accumulated topographically higher than the adjacent sea floor, and in which reef growth was essentially absent" (Playford, 1969, p. 12).
Basin. The open-ocean environment, which extended oceanward from the platform margin, where the water was deeper than on platform, bank, or reef (compare with Cook, 1972).

Basin-slope. The geographically narrow area of steeper slopes transitional between the relatively level platform and basin environments.

Carbonate classification. The carbonate classification used in this report was proposed by Dunham (1962) on the basis of primary depositional textures.

Platform. A geographically widespread mosaic of shallow-subtidal, intertidal, and supratidal carbonate environments. The depositional products of platform environments are areally extensive, sheetlike carbonate bodies, which may form a foundation for areally restricted carbonate banks and reefs (compare with Cook, 1972).

Progradation. The building out of shallow-water environments into deeper water environments, thus leading to upward shoaling in formerly deep environments. Progradation may occur during marine regression and when sediment accumulation exceeds substrate subsidence.

Proximal and *distal.* In this paper these terms refer to the position of a basinal depositional site relative to the platform margin. Proximal environments are relatively closer to the platform margin than distal environments, as determined by direct interfingering relations or by indirect petrographic criteria. The most distal environments discussed in this paper are less than 24 km (15 mi) from the platform margin.

Regression. Offlap of the sea, with the result that offshore environments are replaced by progressively more onshore environments.

Transgression. Onlap of the sea, with the result that onshore environments are replaced by progressively more offshore environments.

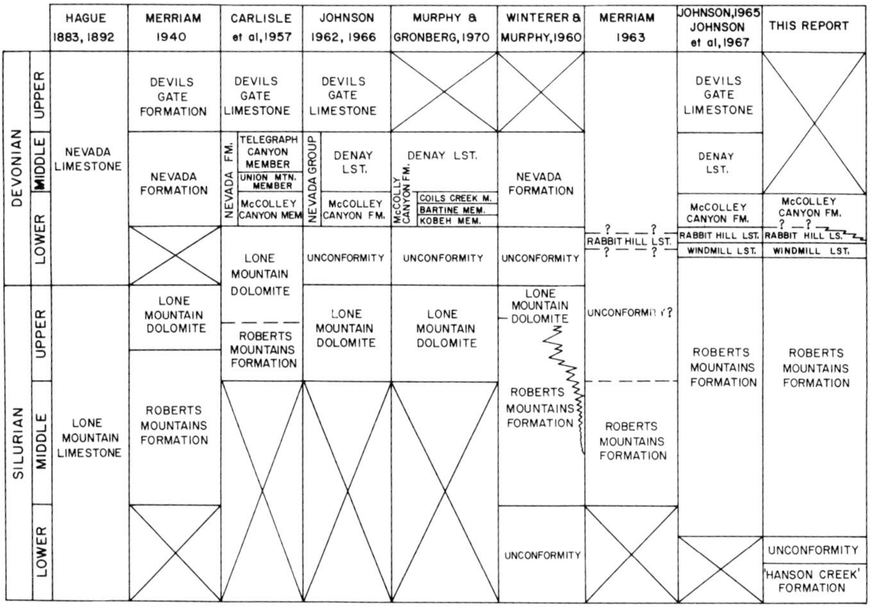

Figure 2. **Diagram summarizing evolution of Silurian-Devonian stratigraphic nomenclature in central Nevada. Diagonal lines indicate time-rock intervals not specifically discussed in the cited paper.**

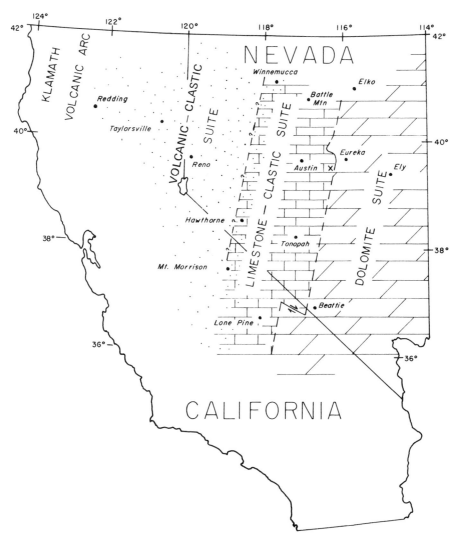

Figure 3. Palinspastic map showing distribution of regional lithofacies in Cordilleran geosyncline of Nevada and California during latest Silurian–earliest Devonian (Pridolian-Lochkovian) time. X = location of Copenhagen Canyon.

Regional Setting

Middle Paleozoic rocks in Nevada and western Utah were deposited in the Cordilleran geosyncline, a continental margin structure representing a "zone of subsidence that extended from the craton to the open ocean" (Roberts, 1972, p. 1994). Cordilleran Silurian and Lower Devonian rocks represent three distinct regional lithofacies (Fig. 3). These time-equivalent facies reflect geographic subdivision of the geosyncline into three tectonic and sedimentary regimes within which distinctly different assemblages of sediment accumulated: (1) an eastern dolomite suite (as used by Berry and Boucot, 1970) deposited on a broad, shallow-subtidal, intertidal, and supratidal carbonate platform; (2) a limestone-clastic suite deposited within a deep-subtidal basin and basin-slope regime oceanward from the carbonate platform margin; and (3) a basinal volcanic-clastic suite deposited oceanward from the limestone-clastic suite.

The facies boundary between the western volcanic-clastic suite and the limestone-clastic suite is poorly understood because regional thrust faulting has obscured their actual paleogeographic relations. Autochthonous volcanic-clastic suite rocks are unknown. Reported occurrences of these rocks consist of allochthonous sheets transported eastward over the limestone-clastic and dolomite suites along low-angle thrust faults (Merriam and Anderson, 1942; Roberts and others, 1958; Roberts, 1964; Gilluly and Gates, 1965; Gilluly and Masursky, 1965; Smith and Ketner, 1968). The volcanic-clastic–limestone-clastic facies boundary is thought to be gradational, and it is always inferred west (oceanward) of known localities where the two facies are tectonically juxtaposed (Roberts, 1964, 1968, 1972; Roberts and others, 1958). Large-scale eastward overthrusting of volcanic-clastic suite rock occurred along the Roberts Mountains thrust and related structures during the Late Devonian–Early Mississippian Antler orogeny (Roberts, 1964, 1968, 1972; Smith and Ketner, 1968; Johnson, 1971; Burchfiel and Davis, 1972; Poole, 1974).

The gradational facies boundary between the eastern dolomite suite and the limestone-clastic suite is better understood. Winterer and Murphy (1960) provided initial documentation of this facies boundary. In the Roberts Mountains they described a Silurian–Lower Devonian limestone-dolomite facies couplet that they interpreted as a local forereef-reef couplet. Cook (1965) described Silurian rocks that accumulated in basinal environments in the Hot Creek Range 129 km (80 mi) south of the Roberts Mountains. The subject of this paper is Silurian–Lower Devonian basin and basin-slope deposits at Copenhagen Canyon that also accumulated adjacent to an eastern shoal-water carbonate platform.

Berry and Boucot (1970, Pl. 1) traced the Silurian facies boundary between the limestone-clastic and dolomite suites throughout the Cordilleran geosyncline. The long-continued pres-

7

ence and geographic extent of the middle Paleozoic facies boundary indicate that it represents a major environmental boundary within the Nevada portion of the Cordilleran geosyncline (Berry and Boucot, 1970; Johnson, 1971; Johnson and others, 1973). We speculate that during Silurian and Early Devonian time the limestone-dolomite facies boundary represented a shelf-slope break that separated a broad inner shelf from a basinal, ensialic outer-shelf regime extending oceanward from the inner-shelf margin. In this proposed model (Fig. 3), the shallow-subtidal, intertidal, and supratidal dolomite suite accumulated on the inner shelf, while the basinal limestone-clastic suite accumulated within the marginal outer-shelf regime. Burchfiel and Davis (1972), Poole (1974), and Churkin (1974) have suggested that allochthonous volcanic-clastic suite rocks in the upper plate of the Roberts Mountains thrust represent ensimatic interarc basin deposits (as used by Karig, 1970, 1971, 1972). This marginal ocean-basin regime extended oceanward from the limestone-clastic suite. A volcanic arc flanked the oceanward side of the interarc basin regime, and this arc provided much of the volcaniclastic and terrigenous sediment in the volcanic-clastic and limestone-clastic suites.

Silurian–Lower Devonian Geology of Copenhagen Canyon

Approximately 762 m (2,500 ft) of Silurian and Lower Devonian beds belonging to the limestone-clastic suite are exposed in the Copenhagen Canyon area (Fig. 4). Three formations are recognized in the Silurian-Devonian sequence (Fig. 5); in ascending order, these are the Roberts Mountains Formation, the Windmill Limestone, and the Rabbit Hill Limestone. The rock units are exposed in a westward-dipping homoclinal block that is moderately deformed by north- and east-trending, high-angle normal and reverse faults (Fig. 4). Ordovician limestone-clastic and volcanic-clastic suite rocks have been thrust over the autochthonous limestone-clastic suite (Bortz, 1959; Merriam, 1963).

ROBERTS MOUNTAINS FORMATION

The name Roberts Mountains Formation was first applied by Merriam (1940, p. 11–12) to a 579-m- (1,900-ft) thick sequence of silty, argillaceous limestone and bioclastic limestone in the Roberts Mountains. The Roberts Mountains Formation is 480 m (1,575 ft) thick at Copenhagen Canyon as measured in section I (Fig. 4).

Lithology. The Roberts Mountains Formation at Copenhagen Canyon consists predominantly of very fine-grained, laminated lime mudstone and wackestone interbedded with subordinate, thick-bedded, graded and nongraded skeletal packstone and grainstone (allodapic limestone; Fig. 6a). The lime mudstone and wackestone weather recessively into gentle, flagstone-covered slopes, while the allodapic limestone forms ledges.

The lime mudstone and wackestone beds are dark gray and black on fresh break and generally weather to light grey and tan gray; weathered strata with high terrigenous content are pale orange or yellow tan. The sediments are generally evenly laminated, although some mudstone and wackestone intervals are thin bedded and massive. In thin and polished section (Figs. 7, 8) the parallel laminae are ½ to 10 mm thick on average, and they are defined by alternations of laminae with high and low allochem percentages or, less commonly, by faint micrograding of fine sand-, silt-, and clay-size sediment. Allochems constitute between 5 and 40 percent of the rock and are supported by a lime-mud and clay matrix (<0.025 mm). Allochems are silt to fine sand size (0.025 to 0.25 mm) and consist of angular to subrounded calcareous skeletal grains, quartz grains, and calcareous grains of unknown origin (Fig. 8). Detrital quartz constitutes up to 20 percent of some wackestone laminae. Elongate allochems

9

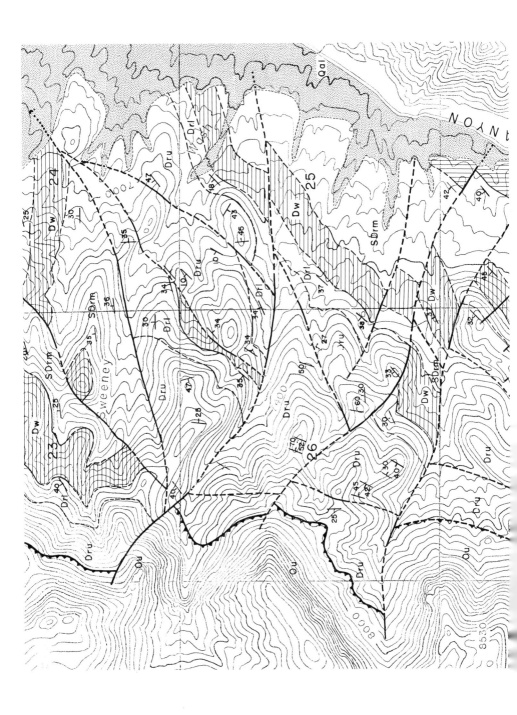

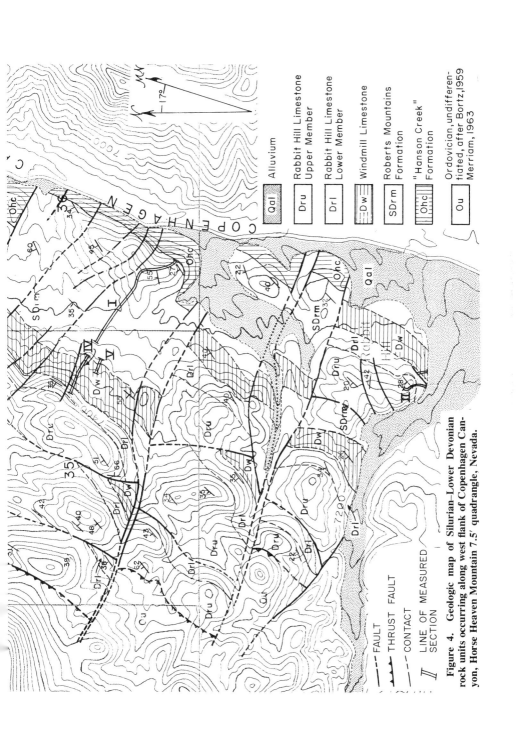

Figure 4. Geologic map of Silurian–Lower Devonian rock units occurring along west flank of Copenhagen Canyon, Horse Heaven Mountain 7.5' quadrangle, Nevada.

Qal — Alluvium

Dru — Rabbit Hill Limestone Upper Member

Drl — Rabbit Hill Limestone Lower Member

Dw — Windmill Limestone

SDrm — Roberts Mountains Formation

Ohc — "Hanson Creek" Formation

Ou — Ordovician, undifferentiated, after Bortz, 1959 Merriam, 1963

----- FAULT

⌐⌐⌐ THRUST FAULT

--- CONTACT

II LINE OF MEASURED SECTION

are oriented parallel to bedding planes in thin and polished section (Fig. 8), but the thin-bedded and laminated sedimentation units appear internally massive in outcrop. Graptolites are common through much of the mudstone-wackestone sequence.

Allodapic packstone and grainstone interbedded with the lime mudstone-wackestone sequence constitute less than 5 percent of the Roberts Mountains Formation, although these beds may be numerous in particular stratigraphic intervals. The allodapic beds are 30 to 40 cm thick on average and are light gray on both fresh and weathered surfaces. The allodapic strata are parallel bedded; the lower contact of each bed is sharp, and the upper contact is generally gradational with overlying lime mudstone, although upper contacts are occasionally sharp. A typical allodapic sedimentation unit consists of a lower skeletal packstone-grainstone interval grading into an upper mudstone-wackestone interval. The packstone-grainstone interval shows normal grading upward from granule- and pebble-size carbonate particles to fine calcareous sand (Fig. 6b). The mudstone-wackestone interval consists of either well-sorted and massive calcareous silt and mud or laminated sand-, silt-, and clay-size calcareous sediment (Fig. 6b). In thin and polished section the allodapic sand is moderately sorted and consists of calcareous skeletal and nonskeletal debris partially supported by grains and partially supported by a lime-mud matrix. Skeletal components consist of fragmental and abraded shoal-water benthic forms, including algae, rugose and tabulate corals, sponge spicules, bryozoa, echinoderms, and disarticulated brachiopods. Nonskeletal components include rounded mudstone intraclasts, pelloids, and uncommon oolites.

Lower Contact. The Roberts Mountains Formation disconformably overlies what has been called the Hanson Creek Formation at Copenhagen Canyon (Bortz, 1959; Merriam, 1963, 1973; Mullens and Poole, 1972). We believe the Hanson Creek rocks are significantly different from those in the type area of the formation, so we use the name "Hanson Creek" in quotation marks.

Figure 9 summarizes stratigraphic relations across the "Hanson Creek"–Roberts Mountains boundary interval. Uppermost "Hanson Creek" strata become progressively siliceous upsection, culminating in a 30- to 35-m sequence of alternating, 2- to 10-cm-thick chert and aphanitic limestone beds. This siliceous interval is overlain by a 2- to 4-m sequence of thin- (1 to 4 cm), medium- (4 to 8 cm), and thick-bedded (8 to 30 cm) bioclastic lime mudstone and wackestone and a few beds of laminated lime mudstone. A lenticular, 0- to 15-cm-thick

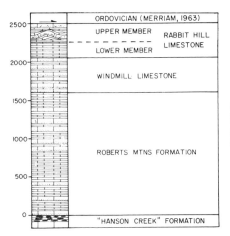

Figure 5. Generalized columnar section showing stratigraphic relations between Silurian–Lower Devonian rock units occurring at Copenhagen Canyon. Thickness measured in feet.

Figure 6.   a. Thick-bedded, allodapic packstone-grainstone bed and enclosing, laminated lime mudstone and wackestone in the Roberts Mountains Formation approximately 52 m (170 ft) above base of formation in Section I.   b. Polished section of allodapic limestone from Roberts Mountains Formation approximately 472 m (1,550 ft) above base of formation in Section V; bar = 1 cm.

bed of phosphatic lime grainstone occurs near the middle of this sequence. An undulatory, irregular bedding surface separates this short interval from a 60- to 100-cm-thick biosparite bed. The lower 30 to 60 cm of this bed contain laminae of sand-size, grain-supported, phosphatic and bioclastic debris. These laminae disappear toward the top of the bed. This phosphatic bed is overlain by laminated, argillaceous lime mudstone and wackestone and allodapic packstone and grainstone typical of the Roberts Mountains Formation. We place the contact between the "Hanson Creek" and Roberts Mountains Formations at the base of the thick phosphatic limestone bed (Fig. 9).

The "Hanson Creek"–Roberts Mountains contact used in this paper differs from stratigraphic usage by previous workers at Copenhagen Canyon, who have included the cherty

Figure 7.   Polished section of lime mudstone-wackestone laminae from Roberts Mountains Formation approximately 5 m (17 ft) above base of formation in Section I; bar = 1 cm.

interval within the Roberts Mountains Formation (Merriam, 1963, 1973; Mullens and Poole, 1972). These previous interpretations were based partly on the fact that a cherty interval was defined as the base of the Roberts Mountains Formation at its type section along Pete Hanson Creek in the Roberts Mountains (Merriam, 1940, p. 11). Previous workers concluded that the cherty unit at Copenhagen Canyon is lithostratigraphically correlative with the basal chert at Pete Hanson Creek. We question this interpretation. We believe that location of the formational contact *above* the siliceous interval and inclusion of the rhythmic limestone-chert sequence *within* the "Hanson Creek" Formation at Copenhagen Canyon is more reasonable, because the siliceous interval is transitional within the upper "Hanson Creek" Formation. Discontinuous secondary chert layers increase in abundance in upper portions of the formation, and a practical mapping contact at this stratigraphic level or at any level within the siliceous interval is not present. In addition, thin- and medium-bedded lime mudstone interbeds in the cherty sequence are lithologically identical to dark-colored lime mudstone beds in lower portions of the "Hanson Creek" Formation. Chert nodules and stringers also occur in these underlying intervals. These features lead us to believe that the secondary chert beds in the rhythmic limestone-chert interval belong within the "Hanson Creek" lithostratigraphic

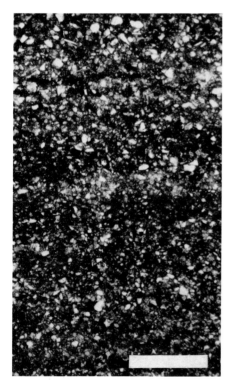

Figure 8. Thin section of lime wackestone laminae from specimen in Figure 7; bar = 1 mm. Silt- and fine sand-size allochems are detrital quartz, calcareous skeletal fragments, and calcareous grains of unknown origin.

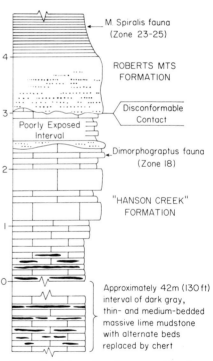

Figure 9. Schematic columnar section showing stratigraphic relations across boundary interval between "Hanson Creek" and Roberts Mountains Formations; units in meters. Composite section from exposures on Martin Ridge and on low hills immediately west of Copenhagen Canyon Road in section 36 (Fig. 4).

sequence and that these cherts have replaced sediments that were deposited during the "Hanson Creek" sedimentary cycle.

The abrupt change from chert-bearing aphanitic limestone and thin-, medium-, and thick-bedded lime mudstone of the "Hanson Creek" Formation to laminated lime mudstone of the Roberts Mountains Formation suggests either (1) an abrupt change in depositional regime or (2) a disconformity and hiatus in sedimentation during which the depositional regime changed. Graptolite biostratigraphic data suggest the latter hypothesis.

Mullens and Poole (1972, p. B28) reported *Climacograptus* cf. *C. rectangularis, Dimorphograptus confertus* cf. var. *swanstoni,* and *Glyptograptus* sp. from near the top of the "Hanson Creek" Formation (Fig. 9) 33 m (102 ft) above the base of their Roberts Mountains Formation). These graptolites characterize the early Llandoverian Zone 18 of Elles and Wood (1901–18). We collected *Monograptus spiralis* and *Retiolites geinitzianus angustidens* 1 m above the base of the Roberts Mountains Formation as recognized in this paper (Fig. 9). These graptolites characterize the late Llandoverian Zones 23 through 25. The sequence of Zone 18 early Llandoverian graptolites followed by Zone 23 through 25 late Llandoverian graptolites within 2 m suggests that a disconformity exists at the top of the "Hanson Creek" Formation.

Placing the formational contact above the siliceous interval is thus supported by both lithologic and biostratigraphic evidence. If the cherty limestone interval is included within the Roberts Mountains Formation at Copenhagen Canyon (in the sense of Merriam, 1963, 1973, and Mullens and Poole, 1972), not only must a practical mapping contact within or at the base of the siliceous interval be demonstrated, but a paraconformity within the Roberts Mountains Formation (in the sense of Mullens and Poole) must be reconciled with an unconformity that occurs at the base of the Roberts Mountains Formation at its type section (Winterer and Murphy, 1960; Berry and Murphy, 1974). We prefer to include the rhythmic limestone-chert interval within the "Hanson Creek" Formation at Copenhagen Canyon. Accordingly, (1) the top of the "Hanson Creek" Formation is early Llandoverian (Zone 18) in age; (2) the base of the Roberts Mountains Formation is late Llandoverian (Zones 23 through 25) in age; (3) the paraconformity at the top of the "Hanson Creek" Formation embraces middle Llandoverian time (Zone 19 through 22, 23, or 24); and (4) the base of the Roberts Mountains Formation at Copenhagen Canyon is not silicified.

WINDMILL LIMESTONE

The Windmill Limestone has not previously been recognized at Copenhagen Canyon. The name Windmill Limestone was applied by Johnson (1965) to a sequence of laminated and bioclastic limestones exposed at Coal Canyon in the northern Simpson Park Range. The Windmill Limestone is approximately 107 m (350 ft) thick at Coal Canyon (Berry and Murphy, 1974); at Copenhagen Canyon it is 139 m (455 ft) thick as measured in Sections IV and V (Fig. 4).

Lithology. Two dominant rock types characterize the Windmill Limestone at Copenhagen Canyon: (1) very fine-grained, laminated, and thin-bedded lime mudstone and wackestone and (2) thick-bedded (0.2 to 1 m) allodapic limestone. Thus, the Windmill Limestone and the underlying Roberts Mountains Formation are lithologically similar, but the two formations are easily distinguished because of the much higher frequency of allodapic limestone in the Windmill. The greater number of ledge-forming allodapic beds causes the

Windmill Limestone to crop out in low, rounded hills above the recessive Roberts Mountains Formation.

Mud-supported and allodapic beds in the Windmill Limestone are lithologically similar to beds in the Roberts Mountains Formation, with the following exceptions: (1) Thin-bedded and internally massive lime mudstone and wackestone beds are as common as evenly laminated, mud-supported intervals, and the massive beds are commonly bioturbated; and (2) Windmill allodapic beds are more abundant, and they are finer grained and better sorted than those in the Roberts Mountains Formation.

Finney (1971) collected *Monograptus birchensis, M. praehercynicus,* and *M. hercynicus* in the Windmill Limestone. The Windmill at Copenhagen Canyon is thus correlative with the lower through upper Lochkovian Stage.

The Windmill Limestone is lithologically similar to the McMonnigal Limestone, a Lower Devonian (Helderbergian) rock unit exposed at Ikes Canyon in the Toquima Range (about 42 km west-southwest of Copenhagen Canyon). Kay and Crawford (1964, p. 440) introduced the name McMonnigal Limestone for "calcisiltites with . . . coquinal brachiopod and coralline beds" that overlie the Masket Shale (= Roberts Mountains Formation). The coquinal brachiopod-coral beds of Kay and Crawford (the massive, 30- to 60-cm-thick limestone beds of McKee and others, 1972, p. 1564) are skeletal lime packstone and grainstone beds that have sharp contacts with underlying laminated and thin-bedded lime mudstone. The well-preserved skeletal debris is commonly normally graded, particularly in the lower part of the McMonnigal, although the packstone-grainstone beds are generally internally massive in upper portions of the formation (E. H. McKee, 1974, personal commun. to Matti). Grain-supported beds predominate over mud-supported beds in most of the formation (McKee and others, 1972, Fig. 3). We believe many of the shelly beds in the McMonnigal were deposited as allodapic sheets or debris flows.

The Windmill and McMonnigal Limestones are so similar lithologically that they may eventually be called the same lithostratigraphic unit. McKee and others (1972, p. 1564) suggested that the McMonnigal Limestone lithologically resembles the Rabbit Hill Limestone at Copenhagen Canyon. However, based on our brief study of rocks at Ikes Canyon, we believe the McMonnigal more closely resembles the Windmill Limestone at Copenhagen and Coal Canyons. The upper and lower members of the Rabbit Hill Limestone (described below) are dominated by laminated and thin-bedded lime mudstone and wackestone. The Windmill Limestone is the rock unit that is characterized by abundant, thick allodapic beds, although the upper member of the Rabbit Hill Limestone contains some thick-bedded allodapic limestone.

Lower Contact. The Windmill Limestone conformably overlies the Roberts Mountains Formation at Copenhagen Canyon. The Roberts Mountains–Windmill boundary is transitional. An 8- to 15-m (25- to 50-ft) stratigraphic interval marking the initial appearance of abundant medium- and thick-bedded allodapic limestone was chosen as the Roberts Mountains–Windmill boundary. The interval is easily recognized and mapped in the Copenhagen Canyon area.

RABBIT HILL LIMESTONE

Merriam (1963, p. 42) applied the name "Rabbit Hill Limestone" to type exposures that underlie Rabbit Hill, a low hill west of the juncture between Whiterock and Copenhagen

Canyons (Fig. 4). However, Merriam did not adequately define the Rabbit Hill Limestone because he did not designate or describe its lower contact. He inferred that the Lower Devonian Rabbit Hill Limestone unconformably overlies Silurian beds of the Roberts Mountains Formation (Fig. 2; Merriam, 1963, Fig. 7, p. 38, 42). Faulting and erosion obscure stratigraphic relations between the Rabbit Hill Limestone and younger rock units throughout Copenhagen Canyon. Merriam reported a thickness of 76 m (250 ft) for the Rabbit Hill at his type section (our Section II, Fig. 4); however, we measured 128 m (420 ft) of Rabbit Hill Limestone in Section II when we discovered additional Rabbit Hill beds higher in the section.

Johnson (1965) recognized the Rabbit Hill Limestone at Coal Canyon in the northern Simpson Park Range (Fig. 1), where the Rabbit Hill overlies the Windmill Limestone and underlies the McColley Canyon Formation in a continuous stratigraphic sequence. Johnson pointed out the structural complexities in the Copenhagen Canyon type area of the Rabbit Hill, and he proposed that a standard section for the Rabbit Hill Limestone be restricted to a "definite sequence of beds in the unfaulted Coal Canyon section," thereby lending itself to a "more exact understanding of Lower Devonian stratigraphy" (Johnson, 1965, p. 373). The Coal Canyon sequence is valuable because it demonstrates vertical stratigraphic relations between the Rabbit Hill Limestone and overlying, younger rock units in the limestone-clastic suite (although the Rabbit Hill does not exhibit the abundance of allodapic beds found at Copenhagen Canyon). The Copenhagen Canyon–Rabbit Hill sequence is valuable because its petrography has been studied in detail and because the biostratigraphy of its brachiopod, conodont, and graptolite faunas has been described (Johnson, 1965, 1970, 1974; Clark and Ethington, 1966; Matti, 1971; J. C. Matti and M. A. Murphy, in prep.; Finney, 1971; Berry and Murphy, 1972). Hence, we believe both the Copenhagen Canyon and Coal Canyon sequences provide necessary reference sections for the Rabbit Hill Limestone.

Mapping by Finney (1971) and Matti (1971) has demonstrated that the Rabbit Hill Limestone occurs within continuous stratigraphic sequences both at Merriam's type section and elsewhere in the Copenhagen Canyon area (Fig. 4). As at Coal Canyon, Rabbit Hill strata conformably overlie beds of the Windmill Limestone. Contact between the Rabbit Hill and Windmill Limestones is best observed in Section IV, and the contact between the two formations is defined in this section (Fig. 10). Lithologic features of the Rabbit Hill Limestone are best observed in Section II (on Rabbit Hill), which is designated the standard reference section for the formation. Contact with the Windmill Limestone is concealed by alluvium at the base of Section II, but this contact can be observed in Section IIA, 50 m east of Section II (Fig. 4). An upper contact between the Rabbit Hill Limestone and younger rock units cannot be defined at Copenhagen Canyon because of faulting and erosion.

The present study recognized two informal members of the Rabbit Hill Limestone, designated the lower and upper members.

Lower Member

Thickness. The lower member is 61 m (200 ft) thick as measured in Section II.

Lithology. The lower member of the Rabbit Hill Limestone consists predominantly of very fine-grained, thin-bedded, and laminated lime mudstone, wackestone, and calcareous shale intercalated with distinctive layers of laminated black chert (Figs. 11a, 12a). Thin- and medium-bedded allodapic limestone is subordinate. Outcrops of the lower member weather recessively and are generally masked by talus from the overlying upper member.

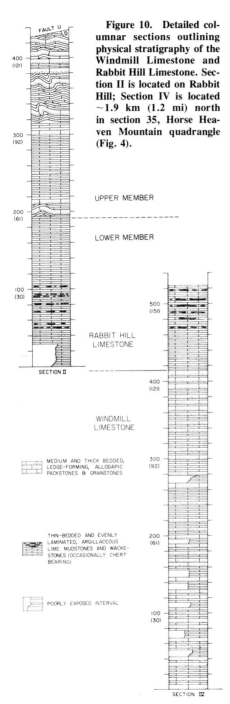

Figure 10. Detailed columnar sections outlining physical stratigraphy of the Windmill Limestone and Rabbit Hill Limestone. Section II is located on Rabbit Hill; Section IV is located ~1.9 km (1.2 mi) north in section 35, Horse Heaven Mountain quadrangle (Fig. 4).

UPPER MEMBER

LOWER MEMBER

RABBIT HILL LIMESTONE

SECTION II

WINDMILL LIMESTONE

MEDIUM AND THICK BEDDED, LEDGE-FORMING, ALLODAPIC PACKSTONES & GRAINSTONES

THIN-BEDDED AND EVENLY LAMINATED, ARGILLACEOUS LIME MUDSTONES AND WACKESTONES (OCCASIONALLY CHERT BEARING)

POORLY EXPOSED INTERVAL

SECTION IV

Figure 11. a. Thin-bedded lime mudstone and wackestone; lower member, Rabbit Hill Limestone 18 to 20 m (60 to 65 ft) above base of formation in Section II. b. Thin section, lime mudstone-wackestone; lower member, Rabbit Hill Limestone ~26 m (85 ft) above base of formation in Section II; bar = 1 mm.

The lime mudstone and wackestone beds are dark gray and gray brown on fresh surfaces and generally weather to tan gray; weathered beds with high terrigenous content are pale orange and yellow tan. The strata are generally thin bedded (1 to 10 cm), although some mudstone-wackestone intervals are evenly laminated. In thin and polished section the parallel laminae are ½ to 10 mm thick on average and are defined mainly by alternation of laminae with high and low allochem percentages. Allochems constitute between 5 and 40 percent of the rock (Fig. 11b) and are supported by a lime-mud and clay matrix (<0.025 mm). Allochems are silt to fine sand size (0.025 to 0.25 mm) and consist of angular to subrounded calcareous skeletal grains, quartz grains, and calcareous grains of unknown origin (Fig. 11b). Detrital quartz constitutes up to 20 percent of some wackestone intervals. Elongate allochems are oriented parallel to bedding planes in thin and polished section (Fig. 11b), but most thin-bedded and laminated sedimentation units appear internally massive in outcrop. Terrigenous content is variable but high throughout the lower member; a distinctive 10- to 15-m stratigraphic interval near the top of the lower member consists of black and gray calcareous shale (Fig. 12a). This stratigraphic interval has yielded the Lower Devonian (Pragian) monograptids *Monograptus thomasi* and *M. yukonensis* (Finney, 1971; Berry and Murphy, 1972).

Beds of black laminated chert between 1 and 15 cm thick are interbedded with the mudstone-wackestone sequence. In thin and polished section the chert has a clastic texture. The grains are silt to coarse sand size and consist of skeletal constituents (including common sponge spicules) and unidentifiable clastic grains. Some of the chert laminae are faintly graded.

Thick-bedded, normally graded limestone is not present in the lower Rabbit Hill member, although thin and medium (1 to 20 cm) allodapic packstone and grainstone beds occur intermittently throughout the interval (Fig. 12b). The allodapic beds are parallel, have sharp lower

Figure 12.   a. Laminated and thin-bedded lime mudstone, wackestone, and calcareous shale; lower member, Rabbit Hill Limestone 38 to 40 m (125 to 130 ft) above base of formation in Section II. b. Thin-bedded allodapic limestone; lower member, Rabbit Hill Limestone ~49 m (160 ft) above base of formation in Section II; note silicified, laminated sedimentation units.

contacts, and have both gradational and sharp upper contacts. In thin and polished section (Fig. 13), a typical allodapic sedimentation unit consists of several successive graded and nongraded packstone-grainstone divisions. The divisions average 0.1 to 5 cm in thickness and consist of grain-supported skeletal and nonskeletal particles. The clastic grains are silt to very coarse sand size (0.025 to 4 mm) and are moderately to well sorted (Fig. 13b). Skeletal components include fragmental algal, echinoderm, coral, bryozoan, brachiopod, and trilobite debris. Nonskeletal components include unidentifiable calcareous grains, pelloids, and uncommon oolites. Cross-stratification of packstone-grainstone laminae occurs in many allodapic intervals (Fig. 13a).

Lower Contact.   The lower member of the Rabbit Hill Limestone conformably overlies the Windmill Limestone (Fig. 10). The Windmill mudstone-wackestone sequence passes essentially unchanged into similar lower Rabbit Hill lime mudstone and wackestone, but the lower Rabbit Hill member is easily distinguished from the Windmill Limestone by its black laminated chert and lack of thick-bedded allodapic sediments. The Windmill–Rabbit Hill contact is drawn at the highest occurrence of thick-bedded allodapic limestone, which characterizes the Windmill Limestone. This contact is defined in Section IV (Figs. 4, 10).

Upper Member

Thickness.   The upper member of the Rabbit Hill Limestone is 67 m (220 ft) thick in Section II (Fig. 10).

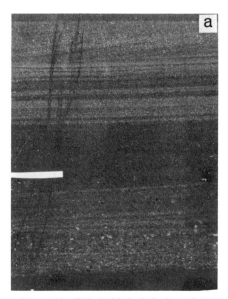 

Figure 13.   Thin-bedded allodapic packstone and grainstone; lower member, Rabbit Hill Limestone ~44 m (145 ft) above base of formation in Section II. a. Polished section showing coarse-grained, faintly graded division, laminated division, and cross-laminated division; bar = 1 cm. b. Thin section of faintly graded, sand-size skeletal debris at base of allodapic grainstone in Figure 13a; bar = 1 mm. Note undulatory but sharp contact with underlying lime mudstone and wackestone.

Lithology. The upper member of the Rabbit Hill Limestone consists predominantly of very fine-grained lime mudstone and wackestone interbedded with subordinate, thick-bedded allodapic limestone. Allodapic beds become more abundant toward the top of the upper member. Upper-member beds have been extensively disturbed by intraformational folding.

Upper Rabbit Hill lime mudstone and wackestone beds are dark and medium gray on fresh break and weather to tan gray. The rocks are either thin and medium bedded (1 to 20 cm) or laminated. In thin and polished section (Fig. 14), mud-supported sediment in the upper member of the Rabbit Hill Limestone is petrographically similar to mudstone and wackestone sediment in the lower member. One major difference is that allochems are so abundant in some wackestone laminae that partially grain-supported packstone textures are developed (Fig. 14). These wackestone-packstone laminae are commonly disturbed by sedimentary boudinage and microfolding (Figs. 14, 19b). Horizontal infaunal burrows occur frequently in the mud-supported sediment.

Thick-bedded (½ to 1 m), graded and nongraded, skeletal packstone and grainstone strata are interbedded with the mudstone-wackestone sequence (Figs. 15, 16). Allodapic strata in the upper Rabbit Hill member contrast strongly with those lower in the section in the following ways. (1) Upper Rabbit Hill allodapic beds are more poorly graded, and their upper contacts are frequently sharp. Typical allodapic beds consist of either (a) a single, poorly graded and poorly sorted packstone-grainstone sedimentation unit with sharp upper and lower contacts (Fig. 15a) or (b) a succession of poorly graded, coarsely laminated packstone-grainstone divisions sharply overlain by lime mudstone (Fig. 15b). (2) Upper Rabbit Hill allo-

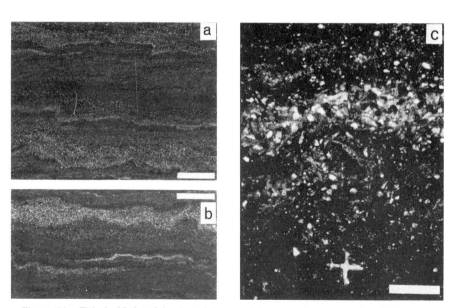

Figure 14.   Thin-bedded and laminated lime mudstone, wackestone, and packstone; upper member, Rabbit Hill Limestone ~75 m (245 ft) above base of formation in Section II. a. Polished section showing microfolding; bar = 1 cm. b. Polished section, different portion of slab in Figure 14a, showing sedimentary boudinage; bar = 1 cm. c. Thin section of specimen in Figures 14a, 14b; bar = 1 mm. Silt-size and fine sand–size grains in packstone lamination include pelloids, skeletal calcite, and detrital quartz. Note sponge spicule near bar scale.

Figure 15. Thick-bedded, skeletal and nonskeletal allodapic packstone-grainstone beds; upper member, Rabbit Hill Limestone in Section II. a. Single, normally graded massive division, showing oriented, disarticulated brachiopod valves; ~61 m (200 ft) above base of formation. b. Faintly graded massive division overlain by coarsely laminated division; arrows point to lime mudstone intraclasts; ~114 m (375 ft) above base of formation.

dapic beds are much coarser grained than underlying allodapic beds. Particles range from cobbles and pebbles to coarse and fine sand-size clastic skeletal grains, shells, and nonskeletal calcareous material (including mudstone intraclasts, pelloids, and uncommon ooliths; Fig. 16b). (3) Upper Rabbit Hill allodapic beds contain a well-preserved, shoal-water benthic fauna (Fig. 16b). The allochthonous faunas include algae, tabulate and rugose corals, entire disarticulated and occasionally articulated brachiopods, echinoderms, bryozoans, gastropods, cephalopods, and trilobites. (4) Many Upper Rabbit Hill allodapic beds exhibit sole structures characteristic of turbidite sedimentation. Load casts and associated flame structures and mud plumes occur at the base of many packstone-grainstone beds (Figs. 16a, 17); pillow-and-ball structures and apparently detached load casts are also present (Fig. 16a).

The entire upper member of the Rabbit Hill Limestone has been disturbed by large- and small-scale intraformational deformation. Outcrop-scale intraformational deformation involves many adjacent sedimentation units and ranges from gentle flexure-slip folds to tightly folded structures, many of which are recumbent or overturned (Fig. 19a). Small-scale deformation at the level of the sedimentation unit includes sedimentary boudinage, microfolds, and convoluted bedding (Figs. 18, 19). These microstructures develop in rocks with mud-supported textures and affect individual sedimentation units and groups of sedimentation units.

Two hypotheses may explain the upper Rabbit Hill intraformational deformation: (1) the structures may represent compressional deformation of Rabbit Hill strata resulting from drag beneath a large-scale thrust fault that carried Ordovician limestone-clastic and volcanic-clastic suite rocks over Rabbit Hill beds (Bortz, 1959; Merriam, 1963); or (2) the association

Figure 16. Thick-bedded, skeletal and nonskeletal packstone and grainstone; upper member, Rabbit Hill Limestone in Section II. a. Coarsely laminated allodapic bed, showing mud plume (1), load cast (2), and apparently detached load cast (or possible allodapic channel; 3); ~120 m (394 ft) above base of formation. b. Polished section of skeletal allodapic bed in Figure 16a; note sharp contact with underlying lime mudstone, and grain-supported skeletal particles including articulated and disarticulated brachiopod valves, crinoid columnals, and rugose corals; bar = 1 cm.

of microstructures and large-scale intraformational folding may represent soft-sediment deformation, the result of upper Rabbit Hill beds slumping down a depositional slope. We favor the latter hypothesis.

The following geometric and kinematic evidence supports the soft-sediment-slump hypothesis. (1) The microfolds and convolute bedding represent passive-flow deformation

Figure 17. Thick-bedded allodapic limestone; upper member, Rabbit Hill Limestone ~116 m (380 ft) above base of formation in Section II. Note load cast and mud plumes.

Figure 18.   Intraformational deformation in upper member of Rabbit Hill Limestone, interpreted as soft sediment slumping down a depositional slope. a. Convolute lamination in lime mudstone and wackestone; note deformation of mud around favositid coral; ~32 m (205 ft) above base of formation in Section II. b. Similar fold in lime mudstone, wackestone, and packstone laminae; ~81 m (265 ft) above base of formation in Section II.

across layer boundaries (Donath and Parker, 1964). The similar folds, convolute bedding, and deformed sedimentary boudins exhibit axial thickening and thinning along their limbs (Figs. 18, 19b). Although thickening and attenuation may occur in brittle materials by mechanical shear and slip, we suggest that thickening and thinning in upper Rabbit Hill microstructures represent ductile flow of incompetent materials. Textural evidence for mechanical shear (textural features such as axial shearing and cleavage and tectonic shear alignment of carbonate grains) is absent; hence, we believe that passive deformation occurred while the sediment was soft. (2) Laminated and thin-bedded mudstone-wackestone intervals between

 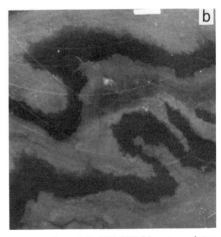

Figure 19.   Intraformational deformation in upper member of Rabbit Hill Limestone, interpreted as soft sediment slumping down a depositional slope. a. Recumbent, overturned, and isoclinal folds 61 to 67 m (200 to 220 ft) above base of formation at Section II; note uncoupling of folds, bedding attenuation, and undisturbed allodapic beds above and below folded interval. b. Microfolding of lime mudstone, wackestone, and packstone laminae; note axial thickening and limb attentuation in the similar folds; polished section, ~75 m (245 ft) above base of formation in Section II; bar = 1 cm.

thick allodapic beds are commonly deformed, whereas the allodapic beds are little disturbed (Fig. 19a). Individual beds may be traced through contorted mudstone intervals bounded by undisturbed allodapic beds and followed out into undeformed intervals bounded by the same undisturbed allodapic beds. This structural relation has two possible origins: (a) ductile mud-supported rocks were slump-deformed between more competent allodapic sediments that remained relatively undisturbed, or (b) in some instances the turbidites may have been deposited on a developing slump. (3) Both large- and small-scale intraformational deformation is confined to the upper Rabbit Hill member; contact between undeformed lower-member beds and folded upper-member beds consistently develops along the same stratigraphic horizon. As there is no evidence for tectonic dislocation along this horizon, we suggest that it reflects a change in the depositional and tectonic environment in which the Rabbit Hill Limestone was accumulating. (4) Qualitative analysis indicates that many observed intraformational folds are recumbent or overturned in the same direction as recumbent and overturned microstructures; this suggests a temporal and genetic correlation between large- and small-scale deformation. (5) Qualitative analysis of measured intraformational fold axes yields random axial trends. If intraformational deformation were the result of tectonic shear beneath a thrust plate, a preferred orientation of fold axes would be expected. Random axial orientation is more readily explained by chaotic gravity-slumping downslope.

Regional thrust faulting and postslump deformation have superimposed a secondary structural overprint on the soft-sediment structures. Drag along high-angle normal and reverse faults has developed local folds and brecciation that can be confused with the soft-sediment structures without careful study. However, we believe that soft-sediment slumping accounts for the major portion of upper Rabbit Hill member intraformational deformation.

Environments of Deposition

Petrographic and stratigraphic relations in the Copenhagen Canyon Silurian–Lower Devonian sequence are similar to those described by Winterer and Murphy (1960) in the Roberts Mountains. At that locality Winterer and Murphy demonstrated that the calcareous Roberts Mountains Formation and the Lone Mountain Dolomite are largely lateral equivalents. They suggested that the limestone-dolomite couplet represents a forereef-reef couplet.

Four basic conclusions are drawn concerning Silurian–Lower Devonian depositional environments at Copenhagen Canyon: (1) The Silurian–Lower Devonian succession is a basinal sequence time-correlative with the Lone Mountain Dolomite and the lowest Nevada Group, shoal-water carbonate bank and platform facies located east of Copenhagen Canyon. (2) The bulk of the Copenhagen Canyon section is a distal sequence deposited farther basinward than time-equivalent proximal sequences in the Roberts Mountains, where the basinal and shoal-water rocks interfinger at the carbonate platform margin. (3) Basinal sediments of the upper Rabbit Hill member were deposited on a slope and were affected by downslope slumping that occurred penecontemporaneously with deposition. (4) Upper Rabbit Hill rocks represent a proximal sequence deposited during basinal progradation of the Nevada Group shoal-water complex.

Depositional environments in the basinal portion of the platform-basin couplet are discussed below. Paleoenvironments on the platform and platform margin and the geometry of the shelf-to-basin transition are not considered. For purposes of the present discussion, the Lone Mountain Dolomite is considered an areally restricted platform margin and (or) bank facies,[1] and the lowest Nevada Group is considered a shallow-subtidal platform facies.

BASINAL ENVIRONMENT: CRITERIA

Evenly laminated, graptolitic, argillaceous lime mudstone and wackestone embracing the Roberts Mountains Formation, Windmill Limestone, and Rabbit Hill Limestone represent

---

[1] The Lone Mountain Dolomite is a pervasively dolomitized facies. Although uncommon dolomite laminae may represent ''primary'' supratidal dolomite, much of the pervasive dolomitization is a secondary (postdepositional) event that has blurred most primary depositional textures. The blurring effect of secondary dolomitization is not restricted to carbonate sediments deposited in platform-margin environments but may extend basinward to affect adjacent basinal mud. Thus, the Lone Mountain Dolomite is not only a lithostratigraphic unit but also a ''metamorphic'' unit, whose basinward extent is determined partly by the distribution of primary depositional environments and partly by the migration of dolomitizing fluids.

marine deposits that accumulated under low-energy conditions. A low-energy depositional regime is suggested by the predominance of mud-supported textures. The parallel laminae are probably the result of low-energy suspension deposition, and they may reflect differential settling of silt- and clay-size carbonate sediment, or episodic or differing rates of supply of these materials. The parallel laminae may also reflect persistent bottom-current activity.

The low-energy depositional milieu at Copenhagen Canyon is interpreted to be a deep-subtidal, below-wave-base, basinal environment. Basinal conditions are suggested by the intercalation of graptolitic lime mudstone and wackestone, which contain no preserved benthic shelly fauna, with graded and nongraded skeletal packstone and grainstone containing a fragmental, shoal-water benthic fauna. The graded and nongraded skeletal packstone and grainstone at Copenhagen Canyon are interpreted as allochthonous carbonate sands (allodapic sands) derived from adjacent contemporaneous shoal-water environments.

A shoal-water, low-energy platform origin might be suggested for the mud-supported rocks at Copenhagen Canyon. Mudstone-wackestone laminite suites with interbedded skeletal packstone and grainstone have been described from Holocene carbonate tidal flats (Illing, 1954; Purdy, 1963a, 1963b; Roehl, 1967; Shinn and others, 1969; Bathurst, 1972; Blatt and others, 1972). Graded and nongraded bioclastic packstone and grainstone in tidal-flat sequences accumulate during storms and anomalously high tidal floods. A distinct suite of primary depositional textures characterizes these intertidal and supratidal carbonate environments; these depositional textures are not present in the Copenhagen Canyon laminite suite. We do not believe the mud-supported sediments at Copenhagen Canyon have a low-energy subtidal, tidal, or supratidal platform origin because these rocks lack an autochthonous benthic shelly fauna, scour-and-fill tidal channels and ripple cross-laminae, dessication features, oolites, laminated dolomite and (or) evaporite suites, fenestrate bird's-eye structures, vertical infaunal burrows, and evidence for an algal or tidal origin for the laminite sequence.

DEPOSITIONAL MECHANISM FOR ALLOCHTHONOUS SAND

Meischner (1964) suggested that allodapic sediments are deposited from turbidity currents. Dott (1963), Sanders (1965), Stauffer (1967), Blatt and others (1972), Cook and others (1972), and Hampton (1972) have pointed out that several mechanisms may be responsible for the mass transport of shoal-water sediments into basinal environments. Sanders (1965) grouped these related mechanisms into the general category of resedimentation. Cook and others (1972, p. 474–479) indicated that the resedimentation mechanism depends largely on the viscosity of the dispersion, and they suggested that three transport mechanisms are possible: turbidity current (turbulent, viscous dispersion), debris flow (laminar, plastic-viscous dispersion), and grain flow (transitional between viscous and plastic-viscous transport; Bagnold, 1956; flowing grain layers of Sanders, 1965; Stauffer, 1967).

We believe that many allodapic sands at Copenhagen Canyon were deposited by turbidity currents, although deposition by grain-flow mechanisms may also have occurred. In interpreting the depositional mechanism for these allodapic sediments, we have concluded that normal grading, lamination, and cross-lamination are primary depositional structures that developed by traction plus fallout (Sanders, 1965) during transport and deposition of the allochthonous debris sheets. These depositional structures probably correspond to Bouma sequences in terrigenous turbidite deposits (Bouma, 1962; Walker, 1965, 1967). However, caution must be observed when comparing primary depositional textures in allochthonous carbonate sands with depositional textures in terrigenous mass-flow deposits, because the hydraulic behavior of skeletal calcite grains (having primary internal porosity) may differ from the hydraulic behavior of clastic silicate grains. Hence, a viscous dispersion of calcareous sediment may behave differently than viscous dispersions of clastic terrigenous sediment, and similar primary depositional textures may or may not occur in calcareous and terrigenous mass-flow deposits, even though they may have accumulated by similar mechanisms.

A turbidity-current origin for allodapic sandstone at Copenhagen Canyon is supported by (1) the presence of sharp lower contacts and both gradational and sharp upper contacts bounding skeletal packstone-grainstone beds; (2) the interpreted presence of complete and truncated Bouma sequences, including normally graded and massive divisions, coarsely and finely laminated divisions, and cross-laminated divisions; and (3) the occurrence of load casts, mud plumes, and apparently detached load casts (pillow-and-ball structures) at the base of many skeletal packstone-grainstone beds. Grain-flow deposition may have produced nongraded allodapic debris sheets with sharp upper contacts, although some top-truncated sequences may reflect bottom-current reworking. Debris-flow deposits (in the sense of Cook and others, 1972) have not been recognized at Copenhagen Canyon.

BASIN-PLATFORM PALEOGEOGRAPHY

The platform margin was located a few kilometers east of Copenhagen Canyon throughout Silurian–Early Devonian time (Figs. 3, 23). The inferred west-east basin-platform geometry is based on indirect petrographic and stratigraphic evidence, however, because the basinal and platform rocks do not interfinger at Copenhagen Canyon as they do in the Roberts Mountains. Cross-lamination in the calcareous turbidites suggests westward sedimentary transport, and the prevailing orientation of recumbent and overturned soft-sediment folds suggests downslope slumping in a westward direction.

The west-east basin-platform paleogeography is confirmed by regional stratigraphic relations. Basinal limestones of the Roberts Mountains Formation, Windmill Limestone, and Rabbit Hill Limestone are absent in outcrops east and northeast of Copenhagen Canyon. The basinal limestone is replaced by time-equivalent dolomite of shoal-water origin.

In the Mahogany Hills and at Lone Mountain, thick sections of Lone Mountain Dolomite stratigraphically overlie thin intervals of the "Roberts Mountains Formation" (Merriam, 1940, 1963). The "Roberts Mountains Formation" is dolomitic at these localities and exhibits primary depositional textures that differ from textures in the calcareous basinal Roberts Mountains Formation at its type section. The dolomitic mudstone and wackestone are not as distinctly laminated, and allodapic packstone and grainstone are not present. Hence, we use the name "Roberts Mountains Formation" in quotation marks (Merriam, 1973, called the Roberts Mountains Formation at Lone Mountain the Lone Mountain Dolomite, Unit 1). The "Roberts Mountains Formation" is gradationally overlain by pervasively dolomitized beds of the Lone Mountain Dolomite, whose relict primary depositional texture includes vugs, relict shell ghosts and impressions, and uncommon algal stromatolites. We believe that the "Roberts Mountains Formation" at Lone Mountain does not represent a basinal facies; it probably represents a low-energy, subtidal platform facies.

Silurian and Lower Devonian limestones are not present in the Eureka district (Nolan and others, 1956) or in the Fish Creek Range (Roberts and others, 1967; Long, 1973). The

Silurian–Lower Devonian interval is occupied by the Lone Mountain Dolomite, Laketown Dolomite, and dolomitic lower Nevada Group. Osmond (1962) suggested that Lower and Middle Devonian dolomites in eastern Nevada and the Eureka district were deposited in shallow-platform environments. Detailed environmental study of the Laketown Dolomite and younger Devonian dolomites in the Fish Creek Range by Long (1973) indicates that these dolomites accumulated in supratidal, intertidal, and shallow-subtidal environments.

Hence, regional stratigraphic relations indicate that a shallow carbonate platform existed east of basinal environments at Copenhagen Canyon throughout Silurian–Early Devonian time. The facies boundary between these two contrasting environments trends generally north-northeast (Fig. 3).

DISTAL ENVIRONMENT

Basinal rocks of the Roberts Mountains Formation, Windmill Limestone, and lower member of the Rabbit Hill Limestone at Copenhagen Canyon represent a distal sequence deposited farther basinward than time-equivalent proximal sequences in the Roberts Mountains and northern Simpson Park Range (Figs. 22, 23). This interpretation is based on (1) comparison of the frequency and petrography of allodapic beds at Copenhagen Canyon with similar attributes in the more northern Silurian–Lower Devonian sequences, and (2) comparison of Copenhagen Canyon allodapic textures with textural criteria for recognizing proximal (near source) and distal (away) turbidite sequences (Walker, 1967).

Proximal environments occur in the Roberts Mountains where the basinal Roberts Mountains Formation interfingers with the Lone Mountain Dolomite platform facies (Winterer and Murphy, 1960). Allochthonous carbonate debris beds containing a well-preserved shoal-water fauna become more abundant, thicker and more irregularly bedded, coarser grained, and more poorly sorted as the platform margin is approached along strike (Winterer and Murphy, 1960; Murphy, 1970). Soft-sediment slumps are common adjacent to the platform margin within both the laminated basinal mud rocks and the allodapic sedimentary rocks. In Early Devonian time the Lone Mountain Dolomite platform complex prograded into the basinal environment (Fig. 22); the Windmill and Rabbit Hill calcareous basinal facies are not present in the Roberts Mountains area. The Windmill and Rabbit Hill facies are present a few kilometers west at Coal Canyon, where allodapic textures in the Windmill Limestone reflect progradation of the Lone Mountain Dolomite in the nearby Roberts Mountains area (Fig. 22). The Windmill allodapic sand and debris-flow beds are proximal, as suggested by their coarse grain size, poor sorting, thick and irregular bedding, and moderately abundant, well-preserved benthic faunas.

In contrast with the northern Silurian–Lower Devonian proximal basinal settings, petrographic and stratigraphic features indicate more distal environments at Copenhagen Canyon. (1) The allochthonous debris beds constitute less than 15 percent of the entire 686-m (2,250-ft) Roberts–Windmill–lower Rabbit Hill interval. Apparently many of the debris sheets that were generated at the eastern shoal-water platform were deposited before they reached the Copenhagen Canyon area. The relative abundance of allodapic beds is strongly dependent on the frequency of the triggering mechanism, however, and allodapic-sediment frequency alone does not confirm a distal depositional setting. (2) The allodapic sandstone is thinner and more parallel bedded at Copenhagen Canyon than in the Roberts Mountains or at Coal Canyon. (3) Allodapic beds at Copenhagen Canyon are finer grained and better sorted than those in the proximal northern localities, and they contain a fragmented and poorly preserved allochthonous shoal-water fauna that contrasts with well-preserved faunas in the proximal northern localities. These compositional and textural features suggest that progressive sedimentary sorting had removed coarser grained, better preserved skeletal and nonskeletal material before the allochthonous debris sheets reached the Copenhagen Canyon area. (4) Soft-sediment slump structures do not occur in the Roberts Mountains Formation, Windmill Limestone, or lower member of the Rabbit Hill Limestone at Copenhagen Canyon. The allodapic beds are undisturbed, and the enclosing basinal mud rocks exhibit planar laminations and thin bedding devoid of intraformational slump structures. Thus, in conjunction with regional facies patterns, the above petrographic features suggest that the Roberts Mountains Formation, Windmill Limestone, and lower member of the Rabbit Hill Limestone at Copenhagen Canyon accumulated on a relatively flat surface basinward from the platform margin.

Walker (1967, p. 25–43) reviewed sedimentary structures and their relation to distal and proximal turbidite environments. Textural characteristics of calcareous turbidites in the Roberts Mountains Formation, Windmill Limestone, and lower member of the Rabbit Hill Limestone at Copenhagen Canyon conform closely to petrographic features characterizing distal turbidite sequences (Walker, 1967, Table 2). These petrographic features include thin graded divisions, fine-grain size, thin and parallel bedding, low sand/mud ratio, well-graded beds, incomplete Bouma sequences, and common lamination.

PROXIMAL, BASIN-SLOPE ENVIRONMENT

The upper member of the Rabbit Hill Limestone is a proximal turbidite sequence deposited on a basin slope. Although recognition of basin slope environments and proximal turbidite environments is based on separate observations, the two environmental reconstructions are compatible and reinforce each other. Together, the slope and proximal environmental interpretations suggest that Copenhagen Canyon was closer to the platform margin during deposition of the upper Rabbit Hill member. Increasing proximity of the slope environment to the platform margin resulted from basinal progradation of the eastern shoal-water carbonate complex during deposition of the upper Rabbit Hill member.

Abundant soft-sediment slump structures provide the basis for our interpretation that the upper Rabbit Hill member was deposited on a slope. As discussed above, large-scale intraformational folds, small-scale microfolds, convolute bedding, and sedimentary boudinage reflect flexural and passive deformation of incompetent, unconsolidated, and semiconsolidated sediments that occurred penecontemporaneously with deposition. The slumps generally do not represent a chaotic mélange of structures, although some slumps have uncoupled along local décollements and travelled some distance downslope. Individual beds can generally be traced through contorted intervals and followed out into undeformed intervals within a single outcrop.

Allodapic textural attributes provide the basis for our interpretation that the upper member of the Rabbit Hill Limestone is a proximal sequence deposited closer to the source of the calcareous turbidites. Not all the textural attributes cited by Walker (1967) for proximal turbidites are manifested by the upper Rabbit Hill allodapic rocks. However, in comparison with allodapic beds lower in the Copenhagen Canyon sequence and in comparison with proximal allodapic beds in the Roberts Mountains and at Coal Canyon, the upper Rabbit

Hill allodapic rocks exhibit proximal fabrics. These include thicker and more irregular bedding, coarser grain size, poorer sorting, common sole marks, high sand/mud ratio, poor grading, and sharp upper contacts. In addition, upper Rabbit Hill allodapic beds contain a well-preserved, shoal-water benthic fauna, whereas underlying distal allodapic beds contain a highly fragmented shoal-water fauna.

Time-Stratigraphic Correlation

Biostratigraphic study of graptolite and conodont faunas at Copenhagen Canyon suggests that the Roberts Mountains Formation is Silurian through Early Devonian (late Llandoverian through early Lochkovian) in age, and the Windmill Limestone is early through late Lochkovian in age (Finney, 1971; Matti, 1971; J.C. Matti and M.A. Murphy, in prep.). We believe the Roberts Mountains Formation and the Windmill Limestone at Copenhagen Canyon are late Llandoverian through Lochkovian time-equivalents of the Lone Mountain Dolomite (Fig. 22). This interpretation is identical to time-stratigraphic correlations proposed for these rock units by Winterer and Murphy (1960), Johnson (1965, 1970, 1971), and Murphy (1970) in the Roberts Mountains district. Hence, we will not discuss this correlation further here.

The following discussion focuses on correlation of the middle Lower Devonian (Pragian) Rabbit Hill Limestone. Refined correlation of the Rabbit Hill Limestone has been difficult because the stratigraphic ranges of Pragian (Helderbergian-Deerparkian) brachiopod and conodont faunas in central Nevada have only recently been worked out in detail. Merriam (1963) published a list of brachiopod and coral species occurring in the Rabbit Hill Limestone at Copenhagen Canyon, and he assigned an Early Devonian (Helderbergian) age to these taxa. However, Merriam did not cite the stratigraphic occurrence or range of these fossils within the formation. Johnson (1965, 1970) studied brachiopods from the Rabbit Hill Limestone at Copenhagen and Coal Canyons, and from the lowest Kobeh Member of the McColley Canyon Formation in the dolomite suite. Johnson's (1970) study of these brachiopods is a definitive taxonomic account of Helderbergian-Deerparkian brachiopods in Nevada. However, his description of the stratigraphic positions of Helderbergian-Deerparkian taxa was only approximate, and stratigraphic relations between these brachiopods and coeval conodont faunas were not adequately known at that time. Johnson (1974) recently clarified the stratigraphic ranges of Helderbergian-Deerparkian brachiopods in the Rabbit Hill Limestone at Copenhagen Canyon based on brachiopod collections obtained by R. A. Flory and us. Matti (1971) and J. C. Matti and M. A. Murphy (in prep.) have studied conodonts interbedded with these brachiopods, and Matti and Murphy are studying conodonts from the lowest Kobeh Member of the McColley Canyon Formation in the dolomite suite. Based on these previous and ongoing biostratigraphic studies, distinct and contrasting stratigraphic models have been proposed for the Rabbit Hill Limestone (Figs. 20, 22).

PREVIOUS STRATIGRAPHIC MODELS

Rock units assigned to the Rabbit Hill Limestone are restricted to the limestone-clastic suite (Fig. 3) where the Rabbit Hill underlies the Nevada Group. The Rabbit Hill interval was

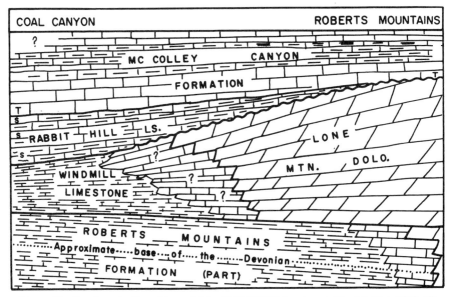

Figure 20. Previous stratigraphic model for Rabbit Hill Limestone (after Johnson, 1970, Fig. 3; T = *Trematospira* Subzone; S = *Spinoplasia* Zone). Time lines are presumably parallel to stratification. Contrast with Figure 22.

formerly thought to be missing in the dolomite suite where the Lone Mountain Dolomite underlies the Nevada Group. Many authors (Nolan and others, 1956; Winterer and Murphy, 1960; Merriam, 1963; Johnson, 1965, 1970) have suggested, on the basis of stratigraphic and structural relations, than an unconformity separates the Lone Mountain Dolomite and the Nevada Group in the dolomite suite (Fig. 2). Winterer and Murphy (1960) and Johnson (1965, 1970) suggested that the Rabbit Hill Limestone is a western depositional equivalent of this unconformity (Fig. 20). This interpretation was made on the basis of brachiopod correlations (discussed below) and because the Rabbit Hill facies is present in the limestone-clastic suite but absent in the dolomite suite. Alternatively, Clark and Ethington (1966) and Matti (1971) proposed that the Rabbit Hill Limestone is coeval with upper portions of the Lone Mountain Dolomite, lower portions of the Nevada Group, and the unconformity between them. Most recently, Johnson (1974) also concluded that uppermost beds of the Rabbit Hill Limestone at Copenhagen Canyon are time-correlative with lower beds of the Nevada Group in the Sulphur Springs Range. Merriam (1973) concluded that the entire Rabbit Hill Limestone is time-correlative with the entire Beacon Peak Dolomite Member of the Nevada Formation (as used by Nolan and others, 1956) throughout central Nevada.

CONODONT BIOSTRATIGRAPHY

Conodont faunas in the Rabbit Hill Limestone at Copenhagen Canyon have been studied by Clark and Ethington (1966), Matti (1971), and J. C. Matti and M. A. Murphy (in prep.).

Conodont faunas from the Kobeh Member of the McColley Canyon Formation in the Sulphur Springs Range have been studied by Klapper at McColley Canyon (Fig. 1; Klapper, 1968; Klapper and others, 1971) and by J. C. Matti, M. A. Murphy, and T. D. Bowden (in prep.) near Telegraph Canyon (SS II, Fig. 1). The biostratigraphy and correlation of pertinent conodont species from these three areas are summarized in Figure 21.

The faunal correlations in Figure 21 suggest that the Rabbit Hill Limestone in the limestone-clastic suite is largely time-equivalent with the lower Kobeh Member in the dolomite suite. Two alternative stratigraphic models are suggested for the Rabbit Hill Limestone (Fig. 22). In both models the upper member of the Rabbit Hill Limestone and youngest beds of the lower member at Copenhagen Canyon are time-correlative with lower portions of the Kobeh in the Sulphur Springs Range. The two models differ only in correlation of basal lower-member beds: (1) basal lower-member beds are partly time-correlative with the upper Lone Mountain Dolomite, the lower Kobeh Member in the Sulphur Springs Range, and the Lone Mountain–Kobeh disconformity (Fig. 22, upper); (2) basal lower-member beds are time-correlative only with the lower Kobeh Member and not with the Lone Mountain Dolomite (Fig. 22, lower). Correlation of the upper Rabbit Hill member and youngest lower-member beds is identical in both models, and we believe this time-correlation is supported by conodont biostratigraphy. Neither stratigraphic model for basal lower-member beds can be demonstrated by existing biostratigraphic data, although either model is consistent with regional depositional relations, and either model fulfills the constraints imposed by the time-space geometry in which the Lone Mountain–Kobeh disconformity and ensuing Kobeh transgressions must have evolved (discussed below). We prefer the lower model depicted in Figure 22 (contrast with Fig. 20).

Merriam (1973, Fig. 4) has suggested that the entire Rabbit Hill Limestone in the limestone-clastic suite is time-correlative with the entire Beacon Peak Member of the Nevada Formation as it occurs in the Eureka district, Roberts Mountains, and southern Sulphur Springs Range. However, Merriam (1973) did not provide biostratigraphic data from the Beacon Peak Member in the southern Sulphur Springs Range (the Beacon Peak is unfossiliferous elsewhere) or from the Rabbit Hill Limestone, whose shelly fossils Merriam groups within a single Helderbergian fauna. Merriam also did not point out that the lower 46 m (150 ft) of his type Rabbit Hill are barren of shelly fossils. For these reasons we cannot evaluate his interpretation that the entire Rabbit Hill Limestone is time-correlative with the entire Beacon Peak Member throughout central Nevada. We suspect that the Beacon Peak in the southern Sulphur Springs Range is probably time-correlative with dolomitic limestone portions of the lower Kobeh Member of the McColley Canyon Formation at McColley Canyon and SS II farther north in the range (Fig. 1). If this is true, Merriam's correlation of part or all of the Beacon Peak Member in the southern Sulphur Springs Range with shell-bearing portions of the Rabbit Hill Limestone is consistent with our stratigraphic model (Fig. 22, lower).

If the time-stratigraphic correlations in Figure 22 (lower) are accurate, as we believe, then the Rabbit Hill Limestone in the limestone-clastic suite is largely time-correlative with the lower Kobeh in the dolomite suite. However, the Kobeh also overlies the Rabbit Hill in the limestone-clastic suite (for example, Coal Canyon). These stratigraphic relations are resolved if the base of the McColley Canyon Formation is diachronous between the dolomite and limestone-clastic suites. Specifically, the base of the Kobeh Member is probably older in the Roberts Mountains and Sulphur Springs Range than it is at Coal Canyon and other localities to the west (Fig. 22, lower; contrast with Fig. 20). This correlation suggests that Kobeh depositional environments migrated diachronously toward the limestone-clastic suite during

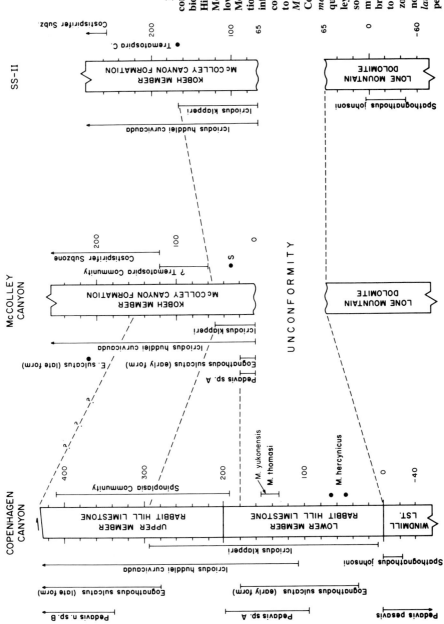

Figure 21. Summary of condont and brachiopod biostratigraphy in Rabbit Hill Limestone, upper Lone Mountain Dolomite, and lower Kobeh Member of McColley Canyon Formation. Dashed lines represent inferred time-stratigraphic correlations. Arrow points to single occurrence of *Monograptus yukonensis* at Copenhagen Canyon. *Trematospira* community is question-marked at McColley Canyon because Johnson (1973, written commun.) cannot assign these brachiopods with certainty to the *Trematospira* Subzone. *Icriodus klapperi* = nomen nudum for *Icriodus latericrescens* ssp. B Klapper (1969).

upper Rabbit Hill time, ultimately replacing Rabbit Hill depositional environments in the limestone-clastic suite (Fig. 22, lower). We interpret this as a shoaling process resulting from basinal progradation of Kobeh platform environments (discussed under depositional environments).

BRACHIOPOD BIOSTRATIGRAPHY

The stratigraphic models for the Rabbit Hill Limestone based on conodont biostratigraphy differ from an earlier model (Fig. 20) based initially on structural and lithostratigraphic evidence (Winterer and Murphy, 1960) and supported later by brachiopod biostratigraphy (Johnson, 1965, 1970, 1974).

Johnson (1970) proposed a biostratigraphic sequence of Lower Devonian brachiopod faunas in central Nevada. He suggested in this zonal sequence that the *Trematospira* Subzone overlies the *Spinoplasia* Zone and that the two brachiopod faunas are entirely superpositional in relation (Johnson, 1965, 1970). The *Spinoplasia* Zone occurs in the Rabbit Hill Limestone and the *Trematospira* Subzone occurs low in the McColley Canyon Formation. At some localities in the dolomite suite, the Kobeh Member and the *Trematospira* Subzone overlie the Lone Mountain Dolomite, and the *Spinoplasia* Zone is missing (Johnson, 1970, Fig. 3). Johnson (1965, 1970) reasonably concluded that the *Spinoplasia* Zone is represented in the dolomite suite by a disconformity between the Lone Mountain Dolomite and the Kobeh Member. The *Spinoplasia* Zone and the Rabbit Hill Limestone in the limestone-clastic suite are western depositional equivalents of this disconformity (Fig. 20).

Johnson's stratigraphic model (Fig. 20) was based on his former belief that the *Spinoplasia* Zone and *Trematospira* Subzone are entirely superpositional in relation. Johnson (1974) revised this belief and suggested that the two assemblage zones partly overlap, although he still maintained that they are largely superpositional. His interpretation is supported by three arguments. (1) The *Spinoplasia* Zone occurs in the Rabbit Hill Limestone, and the *Trematospira* Subzone occurs low in the Kobeh Member of the McColley Canyon Formation. The Kobeh overlies the Rabbit Hill at Coal Canyon. The *Spinoplasia* and *Trematospira* assemblages occur in superpositional sequence at McColley Canyon, where Johnson reported a small collection of *Spinoplasia* Zone shells underlying *Trematospira* Subzone shells in the lower Kobeh (Johnson, 1966, 1973, written commun.). (2) *Spinoplasia* Zone brachiopods are generically correlative with upper Helderbergian and lower Deerparkian faunas in the New York Lower Devonian faunal succession, while *Trematospira* Subzone brachiopods are generically correlative with Deerparkian and Oriskany faunas. (3) The two brachiopod faunas are taxonomically distinct assemblages, despite the fact that many genera and species are common to both faunas. Utilizing these three arguments, Johnson (1965, 1970, 1974) concluded that the *Spinoplasia* and *Trematospira* Zones are largely superpositional, with taxonomic differences between the two assemblages resulting from faunal replacement through time. His recent (1974) suggestion that the two faunas partly overlap in time was based on the occurrence of the Deerparkian genera *Oriskania* and *Acrospirifer* in the uppermost Rabbit Hill Limestone at Copenhagen Canyon.

Conodont biostratigraphy (Fig. 21; Matti, 1971; J. C. Matti and M. A. Murphy, in prep.) suggests greater temporal overlap between the *Spinoplasia* and *Trematospira* assemblages than Johnson (1974) suggested. We believe that the *Spinoplasia* and *Trematospira* assemblages are largely coeval rather than largely superpositional. We suggest a stratigraphic

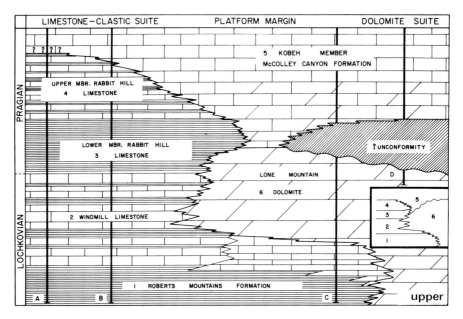

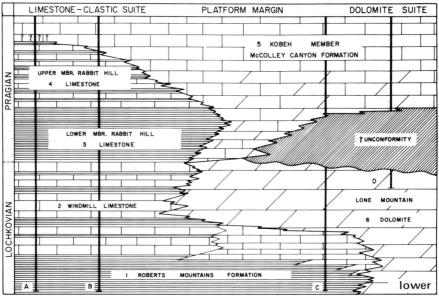

Figure 22.   Alternative stratigraphic models for Rabbit Hill Limestone based on faunal evidence presented in this paper. Neither model can be demonstrated by existing biostratigraphic data. Upper, Rabbit Hill Limestone is partly time-correlative with uppermost Lone Mountain Dolomite. Lower, Rabbit Hill Limestone is entirely time-correlative with lower Kobeh Member of McColley Canyon Formation. With either model, interfingering would occur in the subsurface beneath alluviated valleys where outcrops are nonexistent (for example, Denay Valley, Antelope Valley); contrast with Figure 20. Lettered vertical lines represent the inferred paleogeographic positions of A, Copenhagen Canyon; B, Coal Canyon; C, Roberts Mountains; and D, Sulphur Springs Range. Time lines are parallel to bedding. Inset in upper figure shows relative stratigraphic thicknesses.

model in which the two brachiopod assemblage zones may be largely time-equivalent, facies-controlled benthic communities, each adapted to a different environmental setting (Fig. 21). In this proposed model, the *Trematospira* community was adapted to the shallow-subtidal, carbonate platform Kobeh facies, while the coeval *Spinoplasia* community was adapted to the platform-margin facies. This accounts for the allochthonous occurrence of *Spinoplasia* shells in the basinal Rabbit Hill Limestone, whose allodapic beds were derived from the slope and platform-margin *Spinoplasia* facies rather than from the coeval level-platform *Trematospira* facies behind the platform margin. In this model, taxonomic differences between the two brachiopod faunas can be attributed to differences in coeval community structure rather than to taxonomic change by phyletic or community evolution through time. Taxonomic overlap between the *Trematospira* and *Spinoplasia* communities reflects adaptive flexibility on the part of those taxa that could inhabit both platform and platform-margin facies. Johnson (1974) suggested that the uppermost *Spinoplasia* Zone may partly overlap with the *Trematospira* Subzone and that the *Spinoplasia* and *Trematospira* faunas may in part be coeval benthic communities. However, Johnson still maintained that the two brachiopod faunas are largely superpositional, and he disagreed with our model (Johnson, 1973, written commun.).

If the *Spinoplasia* and *Trematospira* faunas are largely coeval and not superpositional, superposition cannot be invoked to demonstrate that the *Spinoplasia* community (and the Rabbit Hill Limestone) is equivalent to a disconformity merely because the *Spinoplasia* community is largely "missing" between the *Trematospira* community and the Lone Mountain Dolomite. The *Spinoplasia* community is poorly represented in the dolomite suite because it was not adapted to the Kobeh platform environments, as was the coeval *Trematospira* community. Hence, we suggest that brachiopod correlations do not demonstrate that the Rabbit Hill Limestone is entirely a depositional equivalent of the Lone Mountain–Nevada Group disconformity.

REGRESSION, TRANSGRESSION, PROGRADATION, AND THE RABBIT HILL LIMESTONE

Various authors have suggested that a disconformity occurs between the Lone Mountain Dolomite platform complex and the overlying Nevada Group platform complex (Figs. 2, 22; Nolan and others, 1956; Winterer and Murphy, 1960; Osmond, 1962; Merriam, 1963; Johnson, 1965, 1970, 1971). The regional extent of this unconformity is not known; it may be a major regional event affecting the entire dolomite suite platform, or it may be an areally restricted event affecting only a marginal portion of the platform (see Johnson, 1971, Fig. 5). In either case, the unconformity developed initially in the dolomite suite and accompanied a westward regression of dolomite suite platform seas. The regional disconformity has not been recognized in the limestone-clastic suite. Instead, we believe marine offlap of the dolomite suite platform is reflected in the limestone-clastic suite by basinal progradation of the Lone Mountain Dolomite (Fig. 23). Progradation is seen in the Roberts Mountains where the Lone Mountain Dolomite oversteps the Roberts Mountains Formation in a basinal direction in Early Devonian (Lochkovian) time (Fig. 22; Winterer and Murphy, 1960; Johnson, 1965; Murphy, 1970). Regression was followed by marine overlap of the dolomite suite platform during Early Devonian (Pragian) time (Fig. 22, lower). Initial onlap deposits are represented by the basal Nevada Group (McColley Canyon Formation) in the dolomite suite.

What is the stratigraphic position and depositional role of the Rabbit Hill Limestone during these offlap-onlap events and during evolution of the platform margin? We have already proposed temporal relations between the Rabbit Hill and other Lower Devonian rock units involved in the offlap-onlap cycle (Fig. 22, lower), and we have discussed depositional settings for these rock units. We believe these temporal correlations and inferred depositional reconstructions jointly suggest a logical time-space framework relating the Rabbit Hill Limestone to regional offlap and onlap (Fig. 23).

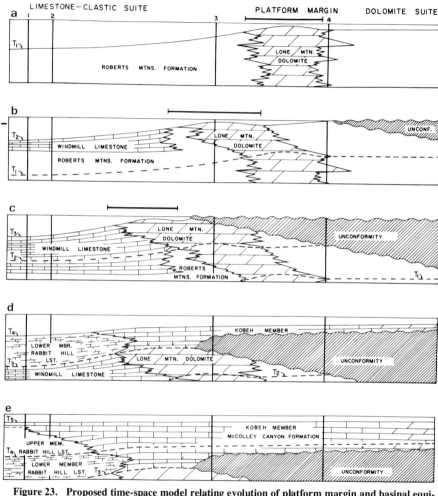

Figure 23.   Proposed time-space model relating evolution of platform margin and basinal equivalents to regional marine offlap and onlap. Each diagram is a time frame during this stratigraphic evolution: a, latest Silurian (Pridolian) time; b, earliest Devonian (early Lochkovian) time; c, Early Devonian (late Lochkovian) time; d, middle Early Devonian (Pragian) time; e, later Pragian time. West-east cross section is ~24 km (~15 mi) across and extends from a few kilometers west of Coal Canyon to the Diamond Range. Numbered vertical lines represent inferred paleogeographic positions of 1, Copenhagen Canyon; 2, Coal Canyon; 3, northern Roberts Mountains; and 4, Sulphur Springs Range. T_1 to T_5 are time lines; otherwise, time lines parallel stratification.

We believe the platform-basin couplet persisted throughout the Lochkovian-Pragian offlap-onlap cycle; that is, the platform and platform-margin facies had basinal equivalents before marine offlap began (Fig. 23a), during the regressive phase (Figs. 23b, 23c), and during the transgressive phase (Figs. 23d, 23e). We believe that basinal progradation of the platform-margin environments was initiated and controlled by marine offlap of the dolomite suite platform. In this model the Lone Mountain platform-margin facies prograded basinward only as far as offlap effects on sea level dictated. Following this interval, marine transgression re-established shallow-platform environments on the Lone Mountain erosional surface. We believe this model has two implications: (1) the prograding Lone Mountain Dolomite had basinal equivalents throughout the extent of platform offlap and during episodes of geographic stability of the platform margin, and (2) lower Nevada Group platform carbonate that onlapped the Lone Mountain erosional surface also had basinal equivalents.

During offlap and progradation, two possible facies relations existed between the platform and its basinal equivalents (Fig. 22, upper and lower): (1) the Windmill Limestone alone is the basinal equivalent of the upper Lone Mountain Dolomite (Fig. 22, lower) or (2) both the Windmill Limestone and basal beds of the Rabbit Hill Limestone are basinal equivalents of this platform complex (Fig. 22, upper).

Sedimentologic and stratigraphic evidence suggests that the Windmill Limestone alone is the basinal equivalent of the prograding platform margin. Winterer and Murphy (1960) and Johnson (1965, 1970) first inferred that the Windmill Limestone at Coal Canyon is time-correlative with the upper Lone Mountain Dolomite (Fig. 20; Winterer and Murphy included the Windmill Limestone within the Roberts Mountains Formation). Winterer and Murphy (1960, p. 135) suggested a near-forereef setting for the Windmill Limestone at Coal Canyon, and they speculated that the Windmill is time-correlative with that portion of the Lone Mountain Dolomite that oversteps the Roberts Mountains Formation in the nearby Roberts Mountains (Fig. 20). Allodapic limestone and debris flows are abundant in the Windmill Limestone at Coal Canyon, and we have already briefly cited petrographic features supporting proximal conditions for these sediments. Allodapic strata dominate the Windmill Limestone at Copenhagen Canyon, although the allochthonous shoal-water faunas are not as well preserved, and proximal textures are not developed as they are at Coal Canyon. Thus, sedimentologic features at both Coal and Copenhagen Canyons support the interpretation that the Windmill Limestone is time-correlative with an actively prograding shoal-water environment that was generating the numerous allochthonous debris sheets present in the Windmill. Conspecific conodonts in the Lone Mountain Dolomite in the Sulphur Springs Range and in the Windmill Limestone support this interpretation (Fig. 21).

Within the offlap-onlap, time-space framework discussed above (Fig. 23), the basinal Rabbit Hill Limestone is also an equivalent of a shallow-platform facies. The lower member of the Rabbit Hill Limestone is either time-equivalent with the prograding Lone Mountain Dolomite (Fig. 22, upper) or is time-equivalent with platform carbonate onlapping the dolomite suite platform (Fig. 22, lower).

Thick bedded allodapic limestone does not occur in the lower Rabbit Hill member at either Copenhagen Canyon or Coal Canyon. Allodapic strata are uncommon and consist of thin-bedded turbidite and grain-flow deposits. The absence of well-developed debris sheets in the lower Rabbit Hill member suggests two depositional models. (1) Basal lower-member beds are proximal basinal equivalents of the Lone Mountain platform-margin facies (Fig. 22, upper), but the platform margin was geographically and sedimentologically stable, thus reducing the potential for generating well-developed allochthonous debris sheets. Platform-margin stability thus accounts for ''distal'' allodapic textures in a proximal lower Rabbit Hill setting. (2) The lower Rabbit Hill Limestone is a distal basinal equivalent of the onlapping Kobeh Member of the Nevada Group (Figs. 22, lower; 23d). In this scheme, distal environments were established during deposition of the lower Rabbit Hill Limestone when the Nevada Group onlapped the dolomite suite platform. Nevada Group onlap would have pushed the shoal-water environments eastward toward the dolomite suite, thus removing the source for well-developed allodapic sediments and producing distal basinal conditions in the limestone-clastic suite during deposition of the lower Rabbit Hill limestone. We prefer this depositional model (Fig. 22, lower) to the upper model in Figure 22.

Johnson (1970, p. 29) suggested that the Rabbit Hill Limestone represents initial deposits of the Helderbergian (Pragian) marine onlap of the dolomite suite platform. In Johnson's model (Fig. 20) the Rabbit Hill Limestone onlaps a disconformable Lone Mountain surface, with subsequent Nevada Group sedimentation beginning virtually isochronously in both the dolomite and limestone-clastic suites. Implicit in this model is that the Rabbit Hill Limestone is a shoal-water carbonate body equivalent in stratigraphic role to a basal transgressive sand. If this is true, the absence of this basal transgressive unit beneath the Nevada Group in the dolomite suite requires explanation. In our model (Figs. 23d, 23e) the Rabbit Hill Limestone does not actively onlap the dolomite suite platform. Rather, distal lower Rabbit Hill muds were being deposited at the same time that basal Nevada Group (basal Kobeh Member) sediments were onlapping the platform. The mud-supported sediment and distal allodapic sediment reflect migration of the shoal-water environments away from the basinal environments.

We believe thick allodapic sand in the upper member of the Rabbit Hill Limestone was deposited in basin-slope environments close to the platform margin. At the same time, conodont biostratigraphy suggests that the upper Rabbit Hill Limestone at Copenhagen Canyon (and probably Coal Canyon) is largely time-correlative with the lower Kobeh Member of the McColley Canyon Formation in the dolomite suite (Fig. 21). The Kobeh overlies the Rabbit Hill at Coal Canyon. These stratigraphic relations suggest that the base of the Nevada Group is diachronous between the dolomite and limestone-clastic suites and that proximal basin-slope conditions in the upper Rabbit Hill member resulted from basinal progradation of the Nevada Group during late Rabbit Hill time (Figs. 22, 23e). In this instance, basinal progradation occurred at the same time as the Nevada Group was actively transgressing the dolomite suite platform. Although there is no preserved record of the prograding platform complex reaching the Copenhagen Canyon area, the Nevada Group probably succeeded the Rabbit Hill Limestone at Copenhagen Canyon as it does at Coal Canyon.

Thus, we believe that during a regional Lochkovian-Pragian offlap-onlap cycle (during which an unconformity developed in the dolomite suite between the Lone Mountain Dolomite and the Nevada Group), the basal Rabbit Hill Limestone was deposited in the limestone-clastic suite while the basal Nevada Group was onlapping the dolomite suite platform (Figs. 22, lower; 23d, 23e). However, we do not discount an alternative model (Fig. 22, upper) in which the Rabbit Hill Limestone was deposited during the waning stages of platform offlap as well as during marine onlap of the dolomite suite platform.

Conclusions

Approximately 762 m (2,500 ft) of Silurian and Lower Devonian beds belonging to the limestone-clastic suite are exposed at Copenhagen Canyon, Nevada. The sequence is conformable and is divided into three formations, including, in ascending order, the Roberts Mountains Formation, the Windmill Limestone, and the Rabbit Hill Limestone (Fig. 5).

Two major lithologies characterize the entire stratigraphic sequence at Copenhagen Canyon. (1) The dominant lithology can be termed the *host lithology*. The host lithology constitutes nearly 90 percent of the total stratal sequence and consists of dark-colored, very fine grained, mud-supported limestone (lime mudstone and wackestone). In outcrop, the lime mudstone and wackestone is evenly laminated (<1 cm) or thin bedded and internally massive. Allochems constitute between 5 and 40 percent of the rock and are supported by a lime-mud and clay matrix (<0.025 mm). Allochems are silt to fine sand size (0.025 to 0.25 mm) and consist of angular to subrounded calcareous skeletal grains, detrital quartz grains, and calcareous grains of unknown origin. Graptolites and tentaculites occur in the mud-supported host rock, but no autochthonous, benthic shelly fossils have been found. (2) The host lithology encloses thin-, medium-, and thick-bedded allodapic lime packstone and grainstone beds that constitute the second major lithology at Copenhagen Canyon. These allodapic limestone beds are calcareous turbidites and grain flows that consist of grain-supported, graded and nongraded, skeletal and nonskeletal calcareous debris derived from shoal-water environments and transported by gravity mass-flow mechanisms into basinal environments. Allodapic constituents are sand through cobble size and consist of fragmented and entire shoal-water shelly fossils.

The entire Silurian–Lower Devonian section at Copenhagen Canyon is thus lithologically very similar throughout its extent. Formal lithostratigraphic units are defined and characterized by (1) the abundance or lack of allodapic beds, (2) the grain size and preservation of allodapic skeletal constituents, and (3) the laminated or thin-bedded and massive character of the lime mudstone and wackestone.

The Roberts Mountains Formation, Windmill Limestone, and Rabbit Hill Limestone had similar depositional histories. The mud-supported host rock in these formations was deposited in low-energy, deep-subtidal, below-wave-base, basin and basin-slope environments. The basinal setting was located marginal to the dolomite suite platform (Fig. 3). Allodapic debris sheets at Copenhagen Canyon were derived from shoal-water environments on the platform margin.

The Roberts Mountains Formation and the Windmill Limestone at Copenhagen Canyon are interpreted as distal, level-bottom sequences. They are western equivalents of the Lone

Mountain Dolomite, a pervasively dolomitized shoal-water carbonate complex that accumu-
lated east of Copenhagen Canyon on the margin of the dolomite suite platform (Fig. 22). The
Roberts Mountains Formation is late Llandoverian through early Lochkovian in age. The
Windmill Limestone is early through late Lochkovian in age.

   An Early Devonian (Lochkovian) disconformity developed in the dolomite suite during
and after deposition of the Lone Mountain Dolomite (Fig. 23). The basal Nevada Group
(Kobeh Member of the McColley Canyon Formation) onlapped this erosional surface during
middle Early Devonian (Pragian) time. The lower member of the Rabbit Hill Limestone represents distal, basinal lime mud that was deposited while the Kobeh Member platform
carbonate onlapped the dolomite suite platform. After the platform was onlapped by the
Kobeh Member, the Kobeh prograded into the limestone-clastic suite basin. Basinal progradation is reflected by soft-sediment-slump structures and thick, proximal allodapic limestone
in the upper member of the Rabbit Hill Limestone. Thus, the upper Rabbit Hill member in the
limestone-clastic suite is time-correlative with the lower Nevada Group in the dolomite suite,
but it was subsequently overlain by the prograding Nevada Group.

   Spinoplasia Zone brachiopods occur in allodapic beds in the upper member of the Rabbit
Hill Limestone. The *Spinoplasia* Zone and the *Trematospira* Subzone were previously
thought to be entirely superpositional in relation. However, we believe the two brachiopod
assemblages are facies-controlled, largely coeval benthic communities (Fig. 20).

References Cited

Bagnold, R. A., 1956, The flow of cohesionless grains in fluid: Royal Soc. London Philos. Trans., ser. A., v. 249, p. 235–297.

Bathurst, R.G.C., 1972, Carbonate sediments and their diagenesis: New York, Americcan Elsevier Pub. Co., Inc., 700 p.

Berry, W.B.N., and Boucot, A. J., 1970, Correlation of the North American Silurian rocks: Geol. Soc. America Spec. Paper 102, 289 p.

Berry, W.B.N., and Murphy, M. A., 1972, Early Devonian graptolites from the Rabbit Hill Limestone in Nevada: Jour. Paleontology, v. 46, p. 261–265.

——1974, Silurian and Devonian graptolites of central Nevada: Univ. California Pubs. Geol. Sci., v. 110 (in press).

Blatt, H., Middleton, G., and Murray, R., 1972, Origin of sedimentary rocks: Englewood Cliffs, N. J., Prentice-Hall, Inc., 634 p.

Bortz, L. C., 1959, Geology of the Copenhagen Canyon area, Nevada [M.S. thesis]: Reno, Univ. Nevada, 56 p.

Bouma, A. H., 1962, Sedimentology of some Flysch deposits: Amsterdam, Elsevier Pub. Co., 168 p.

Burchfiel, B. C., and Davis, G. A., 1972, Structural framework and evolution of the southern part of the Cordilleran orogen, western United States: Am. Jour. Sci., v. 272, p. 97–118.

Carlisle, D., Murphy, M. A., Nelson, C. A., and Winterer, E. L., 1957, Devonian stratigraphy of the Sulphur Springs and Pinyon Ranges, Nevada: Am. Assoc. Petroleum Geologists Bull., v. 41, p. 2175–2191.

Churkin, M., Jr., 1974, Paleozoic marginal ocean basin-volcanic arc systems in the Cordilleran fold belt, in Dott, R. H., Jr., and Shaver, R. H., eds., Modern and ancient geosynclinal sedimentation: Soc. Econ. Paleontologists and Mineralogists Spec. Pub. (in press).

Clark, D. L., and Ethington, R. L., 1966, Conodonts and biostratigraphy of the Lower and Middle Devonian of Nevada and Utah: Jour. Paleontology, v. 40, p. 659–689.

Cook, H. E., 1965, Geology of the southern part of the Hot Creek Range, Nevada [Ph.D. thesis]: Berkeley, Univ. California, Berkeley, 116 p.

——1972, Miette platform evolution in relation to overlying bank ("reef") localization, Upper Devonian, Alberta: Bull. Canadian Petroleum Geology, v. 20, p. 375A–410.

Cook, H. E., McDaniel, P. N., Mountjoy, E. W., and Pray, L. C., 1972, Allochthonous carbonate debris flows at Devonian bank ("reef") margins, Alberta, Canada: Bull. Canadian Petroleum Geology, v. 20, p. 439–497.

Donath, F. A., and Parker, R. B., 1964, Folds and folding: Geol. Soc. America Bull., v. 75, p. 45–62.

Dott, R. H., 1963, Dynamics of subaqueous gravity depositional processes: Am. Assoc. Petroleum Geologists Bull., v. 47, p. 104–128.

Dunham, R. J., 1962, Classification of carbonate rocks according to depositional textures, in Hamm,

W. E., ed., Classification of carbonate rocks — A symposium: Am. Assoc. Petroleum Geologists Mem. 1, p. 108-121.

Elles, G. L., and Wood, E., 1901–18, Monograph of British graptolites: Palaeontographical Soc. London, pt. 1-11, 526 p.

Finney, S. C., 1971, Lower Devonian lithostratigraphy and graptolite biostratigraphy, Copenhagen Canyon, Nevada [M.S. thesis]: Riverside, Univ. California, Riverside.

Gilluly, J., and Gates, O., 1965, Tectonic and igneous geology of the northern Shoshone Range, Nevada: U.S. Geol. Survey Prof. Paper 465, 151 p.

Gilluly, J., and Masursky, H., 1965, Geology of the Cortez quadrangle, Nevada: U.S. Geol. Survey Bull. 1175, 117 p.

Hague, A., 1883, Abstract of report on the geology of the Eureka District, Nevada: U.S. Geol. Survey 3rd Ann. Rept., p. 237–272.

——1892, Geology of the Eureka District, Nevada: U.S. Geol. Survey Mon. 20.

Hampton, M. A., 1972, The role of subaqueous debris flow in generating turbidity currents: Jour. Sed. Petrology, v. 42, p. 775–793.

Illing, L. V., 1954, Bahaman calcareous sands: Am. Assoc. Petroleum Geologists Bull., v. 38, p. 1–95.

Johnson, J. G., 1962, Brachiopod faunas of the Nevada Formation (Devonian) in central Nevada: Jour. Paleontology, v. 36, p. 165–169.

——1965, Lower Devonian stratigraphy and correlation, northern Simpson Park Range, Nevada: Bull. Canadian Petroleum Geology, v. 13, p. 365–381.

——1966, Middle Devonian brachiopods from the Roberts Mountains, central Nevada: Palaeontology, v. 9, p. 152–181.

——1970, Great Basin Lower Devonian Brachiopoda: Geol. Soc. America Mem. 121, 421 p.

——1971, Timing and coordination of orogenic, epeirogenic, and eustatic events: Geol. Soc. America Bull., v. 82, p. 3263–3298.

——1974, *Oriskania* (terebratulid brachiopod) in the Lower Devonian of central Nevada: Jour. Paleontology (in press).

Johnson, J. G., Boucot, A. J., and Murphy, M. A., 1967, Lower Devonian faunal succession in central Nevada, *in* Internat. Symposium on the Devonian System, Calgary, Alberta, 1967 (Proc.): Calgary, Alberta Soc. Petroleum Geologists, v. 2, p. 679–691.

——1973, Pridolian and early Gedinnian age brachiopods from the Roberts Mountains Formation of central Nevada: California Univ. Pubs. Geol. Sci., v. 100, p. 1–75, 31 pl.

Karig, D. E., 1970, Ridges and basins of the Tonga-Kermadec island arc system: Jour. Geophys. Research, v. 75, p. 239–255.

——1971, Origin and development of marginal basins in the western Pacific: Jour. Geophys. Research, v. 76, p. 2542–2561.

——1972, Remnant arcs: Geol. Soc. America Bull., v. 83, p. 1057–1068.

Kay, M., and Crawford, J. P., 1964, Paleozoic facies from the miogeosynclinal to the eugeosynclinal belt in thrust slices, central Nevada: Geol. Soc. America Bull., v. 75, p. 425–454.

Klapper, G., 1968, Lower Devonian conodont succession in central Nevada: Geol. Soc. America, Abs. for 1968, Spec. Paper 121, p. 521–522.

——1969, Lower Devonian conodont sequence, Royal Creek, Yukon Territory, and Devon Island, Canada: Jour. Paleontology, v. 43, p. 1–27.

Klapper, G., Sandberg, C. A., Collinson, C., Huddle, J. W., Orr, R. W., Rickard, L. V., Schumacher, D., Seddon, G., and Uyeno, T. T., 1971, North American Devonian conodont biostratigraphy *in* Sweet, W. C., and Bergström, S. M., eds., Symposium on conodont biostratigraphy: Geol. Soc. America Mem. 127, p. 285–316.

Long, J. F., 1973, Stratigraphy and depositional environments of shoal water carbonate rocks in the Fish Creek Range, central Nevada [M.S. thesis]: Riverside, Univ. California, Riverside, 151 p.

Matti, J. C., 1971, Physical stratigraphy and conodont biostratigraphy of Lower Devonian limestones, Copenhagen Canyon, Nevada [M.S. thesis]: Riverside, Univ. California, Riverside, 148 p.

McKee, E. H., Merriam, C. W., and Berry, W.B.N., 1972, Biostratigraphy and correlation of McMonnigal and Tor limestones, Toquima Range, Nevada: Am. Assoc. Petroleum Geologists Bull., v. 56, p. 1563–1570.

Meischner, K. D., 1964, Allodapische kalke, turbidite in Riff-Nahen sedimentations-Becken, in Bouma, A. H., and Brouwer, A., eds., Turbidites: Amsterdam, Elsevier Pub. Co., p. 156–191.

Merriam, C. W., 1940, Devonian stratigraphy and paleontology of the Roberts Mountains region, Nevada: Geol. Soc. America Spec. Paper 25, 114 p.

——1954, Review of the Silurian-Devonian boundary relations in the Great Basin [abs.]: Geol. Soc. America Bull., v. 65, p. 1284–1285.

——1963, Paleozoic rocks of Antelope Valley, Eureka and Nye Counties, Nevada: U.S. Geol. Survey Prof. Paper 423, 67 p.

——1973, Paleontology and stratigraphy of the Rabbit Hill Limestone and the Lone Mountain Dolomite of central Nevada: U.S. Geol. Survey Prof. Paper 808, 50 p., 12 pl.

Merriam, C. W., and Anderson, C. A., 1942, Reconnaissance survey of the Roberts Mountains, Nevada: Geol. Soc. America Bull., v. 53, p. 1675–1728.

Mullens, T. E., and Poole, F. G., 1972, Quartz-sand–bearing zone and Early Silurian age of upper part of the Hanson Creek Formation in Eureka County, Nevada, in Geological survey research 1972: U.S. Geol. Survey Prof. Paper 800-B, p. B21–B24.

Murphy, M. A., 1970, Preliminary submission for the selection of a type section for the Silurian-Devonian boundary in the Roberts Mountains area of central Nevada, USA: Geol. Newsletter, v. 1970, no. 4, p. 342–360.

Murphy, M. A., and Gronberg, E. C., 1970, Stratigraphy of the lower Nevada Group (Devonian) north and west of Eureka, Nevada: Geol. Soc. America Bull., v. 81, p. 127–136.

Nolan, T. B., Merriam, C. W., and Williams, J. S., 1956, The stratigraphic section in the vicinity of Eureka, Nevada: U.S. Geol. Survey Prof. Paper 276, 77 p.

Osmond, J. C., 1962, Stratigraphy of Devonian Sevy dolomite in Utah and Nevada: Am. Assoc. Petroleum Geologists Bull., v. 46, p. 2033–2056.

Playford, P. E., 1969, Devonian carbonate complexes of Alberta and Western Australia, a comparative study: Western Australia Geol. Survey Rept. 1, 43 p.

Poole, F. G., 1974, Flysch deposits of Antler foreland basin, western United States, in Dickinson, W. R., ed., Tectonics and sedimentation, a symposium: Soc. Econ. Paleontologists and Mineralogists Spec. Pub. (in press).

Purdy, E. G., 1963a, Recent calcium carbonate facies of the Great Bahama Bank — 1, Petrography and reaction groups: Jour. Geology, v. 71, p. 334–355.

——1963b, Recent calcium carbonate facies of the Great Bahama Bank — 2, Sedimentary facies: Jour. Geology, v. 71, p. 472–497.

Roberts, R. J., 1964, Stratigraphy and structure of the Antler Peak quadrangle, Humbolt and Lander Counties, Nevada: U.S. Geol. Survey Prof. Paper 459A, p. 1A–93A.

——1968. Tectonic framework of the Great Basin, in A coast to coast tectonic study of the United States: Missouri Univ. Rolla Tech. Ser., no. 1, p. 101–119.

——1972, Evolution of the Cordilleran fold belt: Geol. Soc. America Bull., v. 83, p. 1989–2004.

Roberts, R. J., Hotz, P. E., Gilluly, J., and Ferguson, H. G., 1958, Paleozoic rocks of north-central Nevada: Am. Assoc. Petroleum Geologists Bull., v. 42, p. 2813–2857.

Roberts, R. J., Montgomery, K. M., and Lehner, R. E., 1967, Geology and mineral resources of Eureka County, Nevada: Nevada Bur. Mines Bull. 64, 152 p.

Roehl, P. O., 1967, Carbonate facies, Williston Basin and Bahamas: Am. Assoc. Petroleum Geologists Bull., v. 51, p. 1979–2032.

Sanders, J. E., 1965, Primary sedimentary structures formed by turbidity currents and related resedimentation mechanisms, in Middleton, G. V., ed., Primary sedimentary structures and their hydrodynamic interpretation: Soc. Econ. Paleontologists and Mineralogists Spec. Pub. 12, p. 192–219.

Shinn, E. A., Lloyd, R. M., and Ginsburg, R. N., 1969, Anatomy of a modern carbonate tidal-flat, Andros Island, Bahamas: Jour. Sed. Petrology, v. 39, p. 1202–1228.

Smith, J. F., and Ketner, K. B., 1968, Devonian and Mississippian rocks and the date of the Roberts Mountains thrust in the Carlin-Pinyon Range area, Nevada: U.S. Geol. Survey Bull. 1251-I, p. 1I–18I.

Stauffer, P. H., 1967, Grain-flow deposits and their implications, Santa Ynez Mountains, California: Jour. Sed. Petrology, v. 37, p. 487–508.

Walker, R. G., 1965, The origin and significance of the internal sedimentary structures of turbidites: Yorkshire Geol. Soc. Proc., v. 35, p. 1–32.

——1967, Turbidite sedimentary structures and their relationship to proximal and distal depositional environments: Jour. Sed. Petrology, v. 37, p. 25–43.

Winterer, E. L., and Murphy, M. A., 1960, Silurian reef complex and associated facies, central Nevada: Jour. Geology, v. 68, p. 117–139.

MANUSCRIPT RECEIVED BY THE SOCIETY NOVEMBER 30, 1973
REVISED MANUSCRIPT RECEIVED APRIL 16, 1974

HILL
REFERENCE
LIBRARY